POLYMER SCIENCE AND TECHNOLOGY

ADVANCES IN CONDUCTING POLYMERS RESEARCH

POLYMER SCIENCE AND TECHNOLOGY

Additional books in this series can be found on Nova's website under the Series tab.

Additional e-books in this series can be found on Nova's website under the e-book tab.

POLYMER SCIENCE AND TECHNOLOGY

ADVANCES IN CONDUCTING POLYMERS RESEARCH

LAURA MICHAELSON
EDITOR

nova publishers
New York

Additional color graphics may be available in the e-book version of this book.

Library of Congress Cataloging-in-Publication Data

Advances in conducting polymers research / editor, Laura Michaelson.
pages cm. -- (Polymer science and technology)
Includes bibliographical references and index.
ISBN 978-1-63463-258-4 (hardcover)
1. Polymers--Electric properties. 2. Conducting polymers. I. Michaelson, Laura, editor.
QD381.9.E38A38 2014
547'.70457--dc23 2014039706

Published by Nova Science Publishers, Inc. † New York

CONTENTS

PREFACE

Conducting polymers (CPs) such as polyaniline (PANI), polypyrrole (PPY), poly(3,4-ethylene dioxythiophene) (PEDOT), and poly(3-hexylthiophene) (P3HT), have been recognized as promising organic semiconductors due to their controllable chemical/electrochemical properties, light weight, low cost, good biocompatibility, facile processability, and adjustable electrical conductivities. This book presents current research in the field of polymers. Topics discussed include resonance raman of polyanilines nanofibers; conducting polymer micro-/nano- structures via template-free method; charge transfer and electrochemical reactions at electrodes modified with pristine and metal-containing films of conducting polymers; and conducting polymer-functionalized carbon nanotubes hybrid nanostructures based bioanalytical sensors.

Chapter 1 – The polyaniline (PANI) and its derivatives are one of the most studied conducting polymers owing to their electrocromic and photoconductivity properties allied with their higher stability in air and easier doping process, as compared to other conducting polymers. These properties turned PANI attractive to use on solar cells, displays, lightweight battery electrodes, electromagnetic shielding devices, anticorrosion coatings and sensors. The recent research efforts are to deal with the control and the enhancement of the bulk properties of PANI, mainly by formation of organized PANI chains in blends, composites and nanofibers. The synthesis of nanostructured PANI, especially as nanofibers, can improve its electrical, thermal and mechanical stabilities. These materials can have an important impact for application in electronic devices and molecular sensors owing their extremely high surface area, synthetic versatility and low-cost. The conventional synthesis of polyaniline, based on the oxidative polymerization of aniline in the presence of a strong acid dopant, typically results in an irregular granular morphology that is accompanied by a very small percentage of nanoscale fibers. However, template-free methods, such as interfacial, seeding and micellar can be employed as different "bottom-up" approaches to obtain pure PANI nanofibers. The possibility to prepare nanostructured PANI by self-assembly with reduced post-synthesis processing warrants further study and application of these materials, especially in the field of electronic nanomaterials. In this chapter this amazing new area of polyaniline nanofibers will be reviewed concerning the state-or-art results of characterization of their structural, electronic and vibrational features. Previous and new results of the spectroscopy of PANI nanofibers and its derivates, obtained by the authors' group, using Resonance Raman will be considered. Special attention will be given in the correlation of PANI nanofibers morphological stabity and their spectroscopic features. The main goal of this work is to contribute in the rationalization of some important results obtained in the open area of PANI nanofibers.

Chapter 2 – This chapter briefly summarizes recent advances in synthesis, characterization of conducting polyaniline (PANI) micro-/nanostructures (e.g., hollow tubes and spheres) via template-free method. The synthesis strategies, self-assembly mechanism and process parameters for the template-free method are discussed. It is found that the morphology (tubes, wires/fibers, and spheres) and size of the PANI micro-/nanostructures can be controlled by adjusting experimental parameters. For example, PANI nanofibers with 10 nm in average diameter have been successfully fabricated. In particular, superhydrophobic films (contact angle can reach up to 148.0^{o}) composed of mono-dispersed PANI nanospheres and oriented-arrays of PANI nanospheres with high crystallinity have also been prepared by this simple and versatile template-free method.

Chapter 3 – Electrospinning is a simple, versatile and efficient method to produce one-by-one continuous ultrathin fibers. Due to low solubility and intrinsic brittleness of conducting polymers (CPs), it is not easy to fabricate CP fibers by direct electrospinning. In the past decade, different strategies have been developed in order to solve this problem and improve electrical conductivity of electrospun CP fibers. This chapter briefly summarizes and reviews three approaches to fabricate CP ultrathin fibers by electrospinning process, including direct electrospinning of CPs into fibers, co-electrospinning of blends of CPs with other spinnable polymers, and template-assisted synthesis using electrospun fibers as templates. In addition, the potential applications of electrospun CP ultrafine fibers in flexible and stretchable electronic devices, field-effect transistors, supercapacitors, neural electrodes and interfaces, etc. have also been discussed.

Chapter 4 – The review is based mainly on the experimental results obtained with electrode systems consisting of different substrates modified by such typical conducting polymers, as polythiophenes and nickel polymer complexes with the Schiff base ligands. The established electrochemical properties of these modified electrodes, as well as the obtained data of their spectroelectrochemical and quartz crystal microbalance studies are discussed in the main part of the review. The performed comparison between these results and those followed from the accepted theory of charge transfer in modified electrodes shows only their qualitative agreement, so that the necessity of improving the existing representations becomes evident. Different methods of syntheses of metal-containing films based on conducting polymers are shortly discussed in connection with the subsequent studies of some electrochemical processes occurring at such composite electrodes. A new approach to treating the polaron conductance of polymer films is proposed. As shown, its inferences significantly differ from the predictions of the existing theory. This permits one to consider the proposed approach as some premise for more detailed studies.

Chapter 5 – Sensors form an integral part of our everyday lives in a wide range of disciplines ranging from detection of environmental toxins, quality control in food and water to healthcare and general safety. Nanomaterials such as carbon nanotubes (CNTs) owing to their small size, high electrical and thermal conductivity, high specific area and superior electronic properties are strong candidates for analyte detection and are thus being increasingly incorporated in sensor architecture. The electrically conducting polymers (CPs) are known to possess numerous features in terms of stability and ease of processing. Their high chemical sensitivity, room temperature operation and tunable charge transport properties has made them ideal for use as transducing elements in chemical sensors. Utilizing the property of surface modification of CNTs, CPs-CNT hybrid structures have been developed by electropolymerization. These hybrid structures exhibit the synergistic benefits of both the materials and allow rapid electron transfer for the fabrication of efficient sensors. This chapter focuses on the synthesis, characterization and applications of conducting polymer-CNTs hybrid nano bio/chemical sensors in various modes of sensor configurations towards sensing gases; volatile organic compounds (VOCs) and biomolecules whose detection and analysis plays a crucial role in environmental pollution control, medical diagnostics and food safety.

In: Advances in Conducting Polymers Research ISBN: 978-1-63463-258-4
Editor: Laura Michaelson

Chapter 1

RESONANCE RAMAN OF POLYANILINES NANOFIBERS

Gustavo Morari do Nascimento*

Universidade Federal do ABC, Centro de Ciências Naturais e Humanas (CCNH)-São Paulo, Santo Bernardo, Brazil

ABSTRACT

The polyaniline (PANI) and its derivatives are one of the most studied conducting polymers owing to their electrocromic and photoconductivity properties allied with their higher stability in air and easier doping process, as compared to other conducting polymers. These properties turned PANI attractive to use on solar cells, displays, lightweight battery electrodes, electromagnetic shielding devices, anticorrosion coatings and sensors. The recent research efforts are to deal with the control and the enhancement of the bulk properties of PANI, mainly by formation of organized PANI chains in blends, composites and nanofibers. The synthesis of nanostructured PANI, especially as nanofibers, can improve its electrical, thermal and mechanical stabilities. These materials can have an important impact for application in electronic devices and molecular sensors owing their extremely high surface area, synthetic versatility and low-cost. The conventional synthesis of polyaniline, based on the oxidative polymerization of aniline

* Corresponding author: Prof. Dr. Gustavo Morari do Nascimento. Universidade Federal do ABC, Centro de Ciências Naturais e Humanas (CCNH)-São Paulo, Santo Bernardo, Brazil. E-mail: gustavo.morari@ufabc.edu.br.

in the presence of a strong acid dopant, typically results in an irregular granular morphology that is accompanied by a very small percentage of nanoscale fibers. However, template-free methods, such as interfacial, seeding and micellar can be employed as different "bottom-up" approaches to obtain pure PANI nanofibers. The possibility to prepare nanostructured PANI by self-assembly with reduced post-synthesis processing warrants further study and application of these materials, especially in the field of electronic nanomaterials. In this chapter this amazing new area of polyaniline nanofibers will be reviewed concerning the state-or-art results of characterization of their structural, electronic and vibrational features. Previous and new results of the spectroscopy of PANI nanofibers and its derivates, obtained by our group, using Resonance Raman will be considered. Special attention will be given in the correlation of PANI nanofibers morphological stabity and their spectroscopic features. The main goal of this work is to contribute in the rationalization of some important results obtained in the open area of PANI nanofibers.

1. General Aspects

1.1. Conducting Polymers

Since the discovery of poly(acetylene) doping process in early 70s [1-6] and posterior investigation of its properties mainly done by Hideki Shirakawa, Alan J. Heeger, and Alan G. MacDiarmid (see Figure 1.1.), the development of the conducting polymer field has continued to accelerate at an unexpectedly rapid rate. This development has been stimulated not only by the fundamental synthetic novelty and importance but mainly because this field is a cross-disciplinary section of investigators- chemists, electrochemists, experimental and theoretical physicists and electronic and electrical engineers, due to the higher potential technological applications.

The doping process [7-14] in polymers is characterized by the passage from an insulating or semiconducting state with low conductivity, typically ranging from 10^{-10} to 10^{-5} Scm^{-1}, to a "metallic" regime (ca. 1-10^{4} Scm^{-1}, see Figure 1.1).

The addition of non-stoichiometric chemical species in quantities commonly low (≤10%), results in dramatic changes in electronic, electrical, magnetic, optical and the structural properties of the polymer. In fact, the dopant chemically reacts with the polymer backbone, and it causes severe disturbance in the crystalline structure of the polymer.

CONDUCTING POLYMERS: GENERAL

σ/ [log/Scm^{-1}]	Polymers	Others
6 5 3	Doped PA Doped PANI	Fe, Cu
0	doped Py doped PTh	Graphite, doped Si
-5	PA	Si, Ti, In
-10	Py (Polypyrrole), PTh (Polythiophene), PANI	Diamond
-15	Nylon	
-20	Teflon	Quartz

doping

NOBEL PRIZE IN CHEMISTRY: 2000

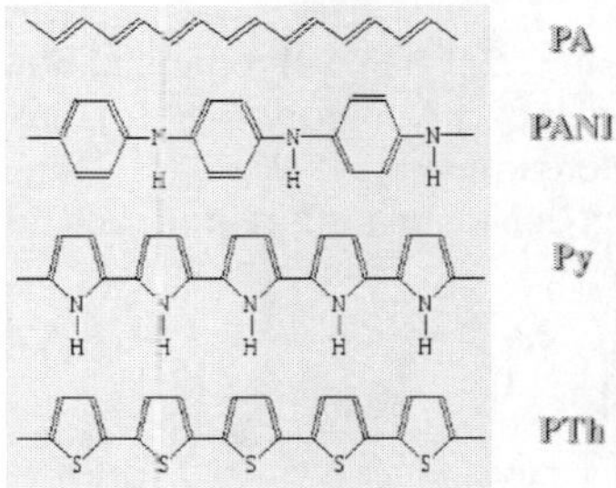

Figure 1.1. The Nobel winners (Hideki Shirakawa, Alan J. Heeger, and Alan G. MacDiarmid) and the schematic representation of the chemical structures of the most common conducting polymers. For comparison purposes the conductivity values for different materials are also displayed in comparison with conducting polymers before and after the doping process.

However, the doping is reversible, and the polymer can return to its original state without major changes in its structure. In the doped state, the presence of counter ions stabilizes the doped state. By adjusting the doping level, it is possible to obtain different values of conductivity, ranging from non-doped insulating state to the highly doped or metallic. All conductive polymers (and their derivatives), for example, among others, may be doped by p (oxidation) or n (reduction) through chemical and/or electrochemical process [6-8] (see Figure 1.2).

1.2. Polyanilines

The doping process can also be characterized by no lose or gain of electrons from external agents. This is the point for Polyanilines, and this process is named internal redox process.

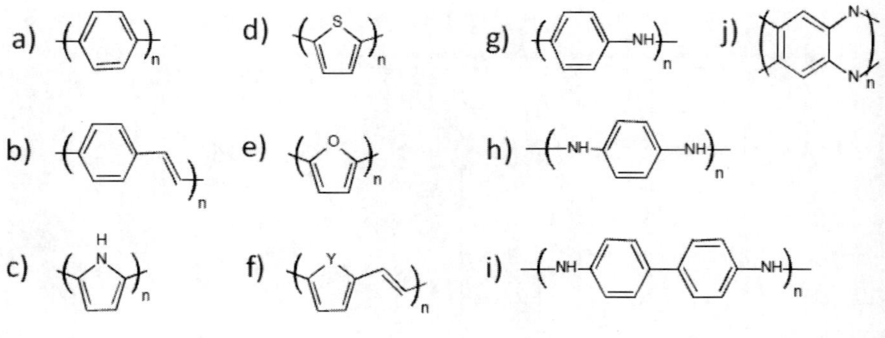

Figure 1.2. Chemical representation of poly (*p*-phenylene) (a), poly(*p*-phenylene-vinylene) (b), poly(pyrrole) (c), poly(thiophene) (d), poly(furan) (e), poly (heteroaromatic vinylene) (f, where Y = NH, NR, S, O), poly(aniline) (g), poly(*p*-phenylenediamine) (h), poly(benzidine) (i), and poly(*o*-phenylenediamine) (i).

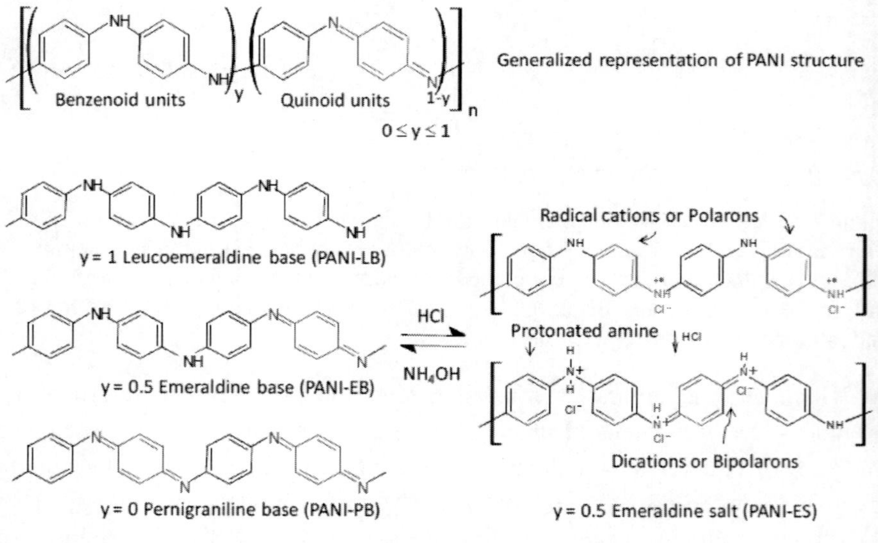

Figure 1.3. Generalized representation of chemical structure of PANI and its most common forms.

For instance, poly(aniline) (PANI) in its insulate emeraldine base form (PANI-EB, the most stable form of PANI) can be converted to the doped form (emeraldine salt form, PANI-ES) by simple protonation with strong acids (see Figure 1.3) [12-14]. By mainly protonation of imine (and sometimes also amine) nitrogens is observed the formation of charged segments or species, as radical cations (polarons) and dications (bipolarons) inside the polymer backbone.

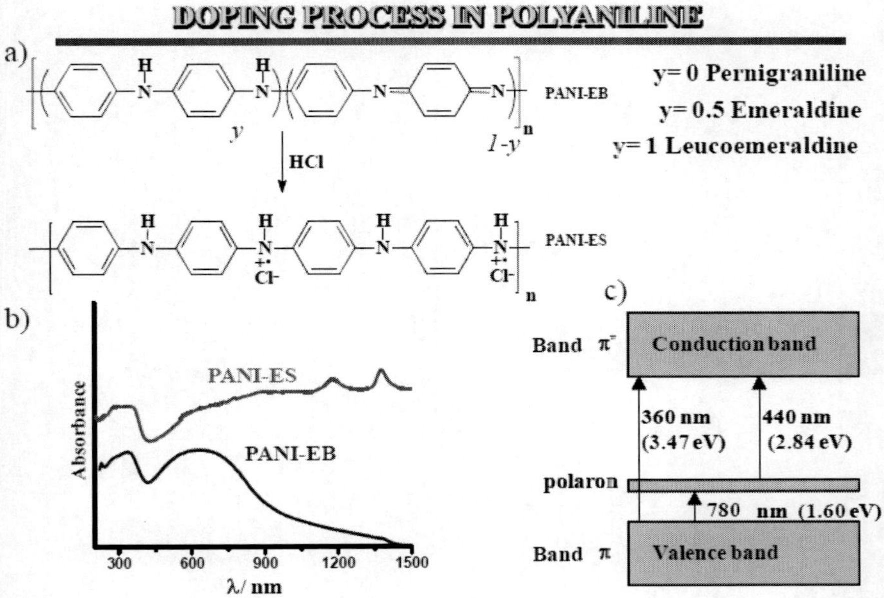

Figure 1.4. Generalized representation of doping process in PANI. a) Main chemical modification, b) UV-vis-NIR changes and c) Electronic levels of PANI-ES form.

The conductivity of the polymer can be increased by more than 10 times, reaching to 3 $S.cm^{-1}$ [12-14]. The doping with protonic acids was also observed later for the poly(heteroaromatic vinylene) [9]. The changing of oxidation and protonation levels in PANI structure can be visualized by monitoring its electronic and/or vibrational spectra. For all oxidation states of PANI the absorption band in the UV region is related to the transition $\pi-\pi^*$ of the benzene ring. After protonation with the formation of doped PANI (PANI-ES, see Figure 1.4, part (a)), it is observed a band at visible-NIR region (1.6 eV or 780 nm, see Figure 1.4, part (b)), which is attributed to a charge transfer from the highest occupied energy level of the benzene ring (HOMO) to the lowest unoccupied energy level of a semi-quinone ring (LUMO), it is characteristic of the doping state and is represented in Figure 1.4, part (c) [12-14]. MacDiarmid et al. [15] studied by UV-vis-NIR the changes that occurs during the protonation of PANI-EB. Figure 1.5 part (1) shows what happens in the UV-vis-NIR spectra of PANI-EB during its protonation with hydrochloric acid. The spectra at pH 6 (A) and pH 4 (B) are identical, but with increase of the acidity of the medium (Spectrum B to G), the band at ca. 2.1 eV (595 nm) shifts to approx 1.6 eV (780 nm), as consequence of the structural distortion of PANI chains with formation of radical cations (polarons).

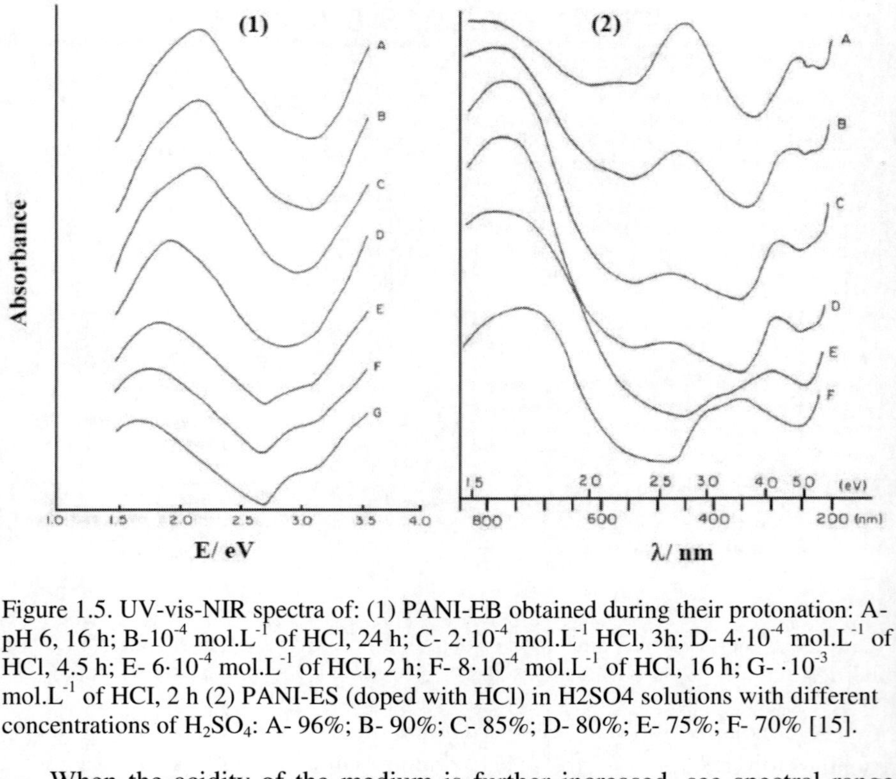

Figure 1.5. UV-vis-NIR spectra of: (1) PANI-EB obtained during their protonation: A- pH 6, 16 h; B-10^{-4} mol.L^{-1} of HCl, 24 h; C- $2{\cdot}10^{-4}$ mol.L^{-1} HCl, 3h; D- $4{\cdot}10^{-4}$ mol.L^{-1} of HCl, 4.5 h; E- $6{\cdot}10^{-4}$ mol.L^{-1} of HCI, 2 h; F- $8{\cdot}10^{-4}$ mol.L^{-1} of HCl, 16 h; G- $\cdot 10^{-3}$ mol.L^{-1} of HCI, 2 h (2) PANI-ES (doped with HCl) in H2SO4 solutions with different concentrations of H_2SO_4: A- 96%; B- 90%; C- 85%; D- 80%; E- 75%; F- 70% [15].

When the acidity of the medium is further increased, see spectral range from F to A, see figure 1.5, part (2), the intensity of the band at ca 3.0 eV (416 nm) decreases for lower pH values. Furthermore, new bands at ca 2.5 eV and 1.6 eV appears. MacDiarmid et al. [15] suggests that this may be related to the transition of radical cations (polaronic segments) to dications (bipolaronic segments, see Figure 1.3) units. Finally, the dissolution study of PANI-CSA in appropriate solvents (usually phenol) takes the appearance of a strong absorption in the NIR region. MacDiarmid suggested the formation of free charge carriers and forming extended polymer chains.

2. Resonance Raman

2.1. General Aspects

Raman spectroscopy is a technique *par excellence* for probing the vibrational frequencies by scattering the incident light, usually in the visible range. In the off-resonance Raman spectroscopy (sometimes called normal Raman spectroscopy) the intensities of the Raman bands are linearly proportional to the intensity of the incident light (I_o, see Figure 2.1), proportional to the fourth power of the wavelength of the scattered light (λ_s^4 or ν_s in wavenumber units, see Figure 2.1), and proportional to the square of the polarizability tensor ($[\alpha]^2$) [16-19]. The situation changes dramatically, when the laser line falls within the region of a permitted electronic transition. The Raman intensities associated with vibrational modes which are tightly coupled or associated with the excited electronic state can suffer a tremendous increase of about 10^5 powers; this is what characterizes the resonance Raman effect. (see Figure 2.1). The mathematical and theoretical backgrounds used to the interpretation of the resonance Raman behavior can be found extensively in the literature [16-19]. Generally, the tensor of polarizability is described as shown in the Figure 2.1. The equation is formed in the numerator part by transition dipole moment integrals between the electronic ground state (g, for the vibrational m or n states) and an excited electronic state (e, for any vibrational v states). The sum is done over all possible (e,v) states. In the denominator part is the difference or sum of the scattered and incident light, added by the dumping factor ($i\Gamma_{ev}$) that contents information about the lifetime of the transition states. The theoretical formalism developed by Albrecht et al. is commonly employed [16-19]. This enormous intensification makes, in principle, the Raman spectrum easy to be acquired.

But, in a state of resonance, a lot of radiation is absorbed, leading to a local heating and frequently can be observed a decomposition of the conducting polymer. Despite of this problem, the RR spectroscopy has been largely used in the study of the different chromophoric units present in polyaniline and others conducting polymers, just by tuning an appropriate laser radiation on an electronic transition of the polymer. This behavior is clear visualized in Figure 2.2, where the PANI spectrum changes dramatically with the laser line used in the Raman measurements.

PANI shows a characteristic Raman bands for each oxidized or protonated form (see Figure 2.2) [20, 21]. The Raman spectrum of fully reduced PANI (applied potential of -100 mV) was identified as being formed by benzenoid rings. In contrast, the intensity of the Raman spectra obtained for PANI at 632.8 nm (E_{laser}= 1.97 eV) increased when PANI was oxidized. At applied potential of +600 mV three Raman bands (1160, 1490 and 1595 cm^{-1}) were identified as characteristics of the quinoid structure of PANI. Figure 2.2 presents the segments of PANI and its characteristic Raman bands at their corresponding exciting radiation [20-23].

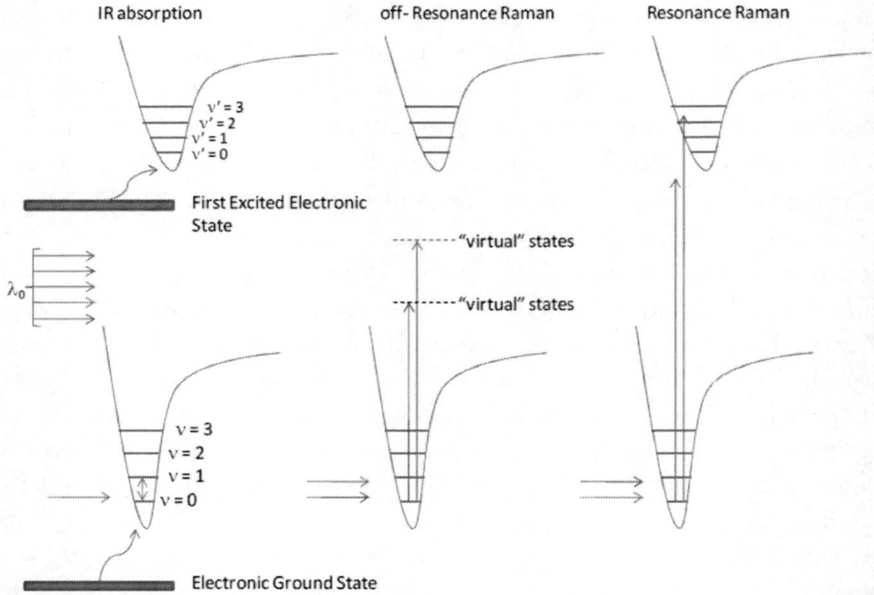

$$I_{Raman} \propto I_o \nu_s^4 \sum_{\rho\sigma} \left|[\alpha_{\rho\sigma}]_{gn,gm}\right|^2$$

$$[\alpha_{\rho\sigma}]_{gn,gm} = \frac{1}{hc}\sum_{ev}\left[\frac{\langle n|[\mu_\rho]_{ge}|v\rangle\langle v|[\mu_\sigma]_{eg}|m\rangle}{\nu_{ev,gm}-\nu_0+i\Gamma_{ev}} + \frac{\langle n|[\mu_\sigma]_{ge}|v\rangle\langle v|[\mu_\rho]_{eg}|m\rangle}{\nu_{ev,gn}+\nu_0+i\Gamma_{ev}}\right]$$

Figure 2.1. Schematic representation of two electronic states (ground and excited) and their respective vibrational levels. The arrows indicated the types of transitions that can be occurred among the different levels. It is important to say that in the case of Raman scattering, if the used laser line (λ_o, or as wave number, represent by ν_o) has energy similar to one electronic transition of the molecule, the signal can be intensified, known as resonance Raman Effect. In the Figure ν_o and ν_s (the scattered frequency is composed by: $\nu_{ev,gm}$ and $\nu_{ev,gn}$, the stokes and anti-stokes components, respectively) are the laser line and the scattered frequencies. It was given the equations that describe the Raman Intensity and also the tensor of polarizability. The equation is formed in the numerator part by transition dipole moment integrals between the electronic ground state (g, for the vibrational m or n states) and an excited electronic state (e, for any vibrational v states). The sum is done over all possible (e,v) states. In the denominator part is the difference or sum of the scattered and incident light, added by the dumping factor ($i\Gamma_{ev}$) that contents information about the lifetime of the transition states.

The PANI-LB is characterized by the vibrational modes of the benzene ring in 1618 and 1181 cm^{-1}, attributed to the νCC and βCH, respectively. The amine group is characterized by νCN stretch at 1220 cm^{-1}. For PANI-PB the βCH band value is at 1157 cm^{-1}, and another characteristic band of PANI-PB is the stretch of C=N bond at 1480 cm^{-1}. Another way to determine the degree of oxidation of PANI [22], consists in determination of the intensities of the bands at about 1500 cm^{-1} for PANI-LB (νCC) and the band around 1600 cm^{-1} for PANI-PB (νC=C) observed in the IR spectra. The intensity ratio between these two bands (I(1600)/I(1500)) is a way to determine qualitatively the degree of oxidation in the chain of PANI.

The Raman studies of PANI-ES suggest the existence of bipolaronic segments (dications or protonated imines) [24]. The presence of these segments was also indicated by UV-vis-NIR data [15] and by EPR [25] The origin of doublet nature of the CN stretch (ca. 1320-1350 cm^{-1}) remains unclear [21]. But, some authors suggested, that [21] the doublet may be associated with the existence of two different conformations of PANI. The Raman study of PANI doped with camphorsulfonic acid (CSA) and dissolved in *m*-cresol [26, 27] revealed a conversion of dications to radical cations. This behavior is associated with changes in the electronic structure, leading to the appearance of new Raman bands and the modifications of others, due to, the high charge delocalization on the polymeric chains [28, 29]. In Figure 2.3 it is seen the spectral change of the Raman spectra of PANI from EB to ES forms and it is clear the decrease of the bands associated to polaronic/bipolaronic units and the increase of the bands associated to neutral and oxidized units of PANI.

The Raman studies of PANI using near-infrared (NIR) laser line is also found [30-33]. The most peculiar feature observed at 1064.0 nm is the presence of a sharp band around 1375 cm^{-1} in PANI-EB spectrum, which was correlated to polaronic segments localized at two benzene rings. On the other hand [30], it was proposed that this band was not correlated with protonated segments but with over-oxidized segments such as those present in PANI-PB. Some controversial aspects about the Raman spectra of PANI at NIR excitation were recently re-examined [33]. The bands from 1324 to 1375 cm^{-1} were associated to νC–N of polarons with different conjugation lengths and with the presence of charged phenazine-like and/or oxazine-like rings in PANI-ES as chemically prepared. The formation of cross-linking structures is associated with the ES form of PANI.

The bands from 1450 to 1500 cm^{-1} in the PANI-EB and PANI-PB spectra were associated with the νC=N mode of the quinoid units having different conjugation lengths.

The thermal behavior of PANI revealed that there is the appearance of intense bands at 574, 1393 and 1643 cm^{-1} in the Raman spectra at 632.8 nm during heating [26, 27, 34].

The same behavior is observed in the poly(diphenylamine) doped with HCSA (PDFA-CSA) during heating [34]. By comparing the results obtained from the thermal monitoring of PANI-CSA and PDFA-CSA, it was possible to assign these bands to the reaction of the polymer with oxygen, with formation of chromophores with oxazine-like rings.

It was also demonstrated that the increase of laser power at 1064.0 nm causes deprotonation of PANI-ES and formation of cross-linking segments having phenazine and/or oxazine-like rings. The formation of cross-linking structures is associated with the ES form of PANI.

The resonance Raman studies of the PANI-CSA [26, 27, 34] treated with *m*-cresol, named secondary doping, revealed that this process causes a conversion of dication to radical cations structures. This behavior is explained by the increase of the band at ca. 1336 cm^{-1}, assigned to νCN of polaronic segments, and the intensity decreases of bands at 1486 cm^{-1} and 1380 cm^{-1}, assigned to νC=N and νC=C vibrational modes of dications segments, respectively, in the Raman spectrum of PANI-CSA treated with *m*-cresol at 632.8 nm laser line.

M. Cochet et al. [28, 29] also investigated this process using resonance Raman spectroscopy. The authors tried to analyze the secondary doping by normal mode coordinates approach, as conclusion the results cannot be solely rationalized by changing in the benzene rings planarity. The secondary doping behavior is also associated to changes in the electronic structure, it leading to the appearance of new Raman bands. The thermal behavior of PANI-CSA was monitored using *in situ* Raman spectroscopy by Da Silva et al. [26, 27] and Do Nascimento et al. [34], and it revealed the appearance of intense bands at 574, 1393 and 1643 cm^{-1}, those are resonant at 632.8 nm laser line.

By comparing these results with similar study of poly(diphenylamine)-CSA (PDPA-CSA) it was possible to assign the bands at ca. 583, 1398 and 1644 cm^{-1} (574, 1393 and 1643 cm^{-1} band values for PANI-CSA) to the polymer reaction with oxygen followed by formation of chromophoric segments with oxazine-like rings (see Figure 2.4).

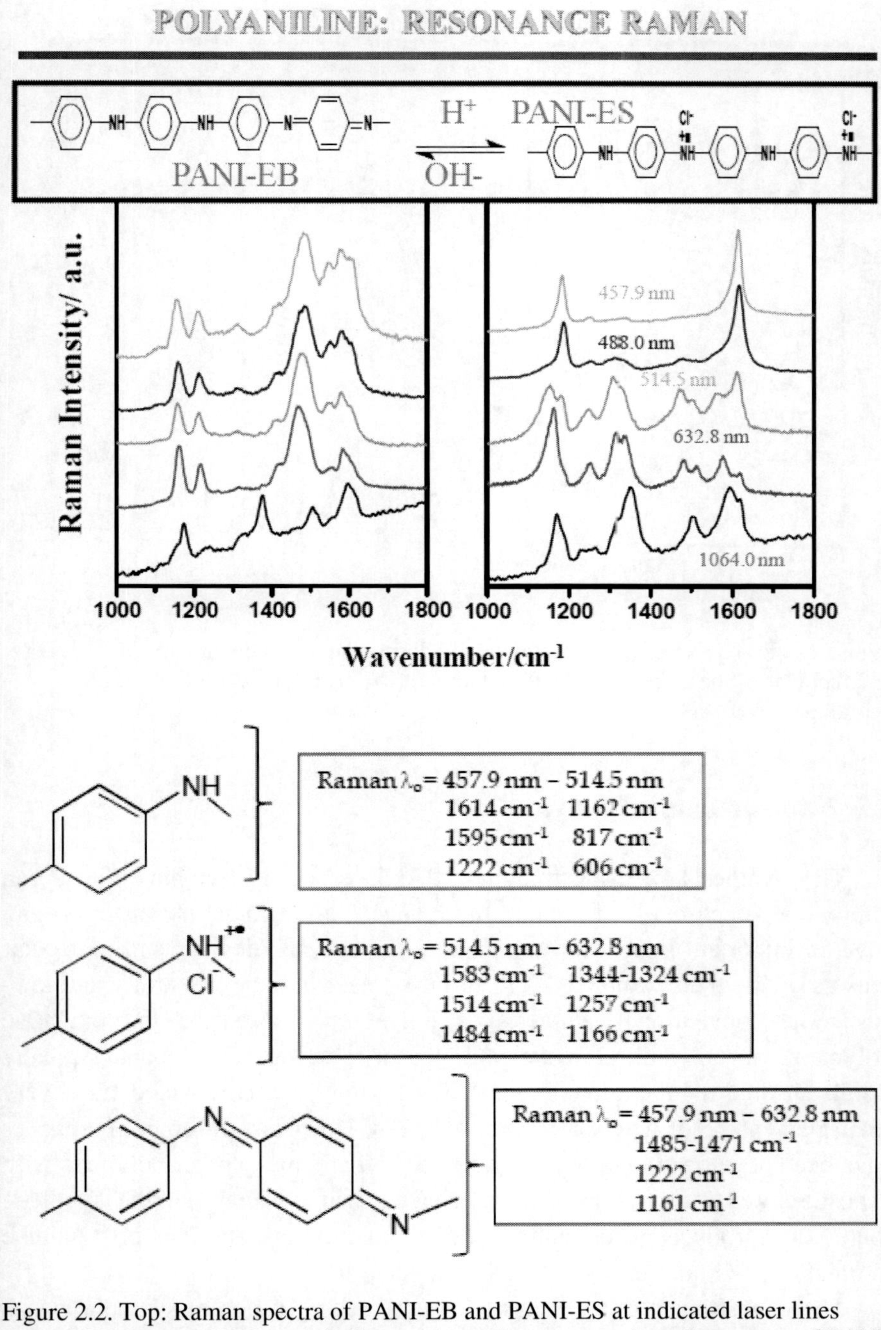

Figure 2.2. Top: Raman spectra of PANI-EB and PANI-ES at indicated laser lines (from 1064.0 nm to 457.9 nm). Bellow: schematic representation of segments of PANI and its characteristic Raman bands at indicated laser lines.

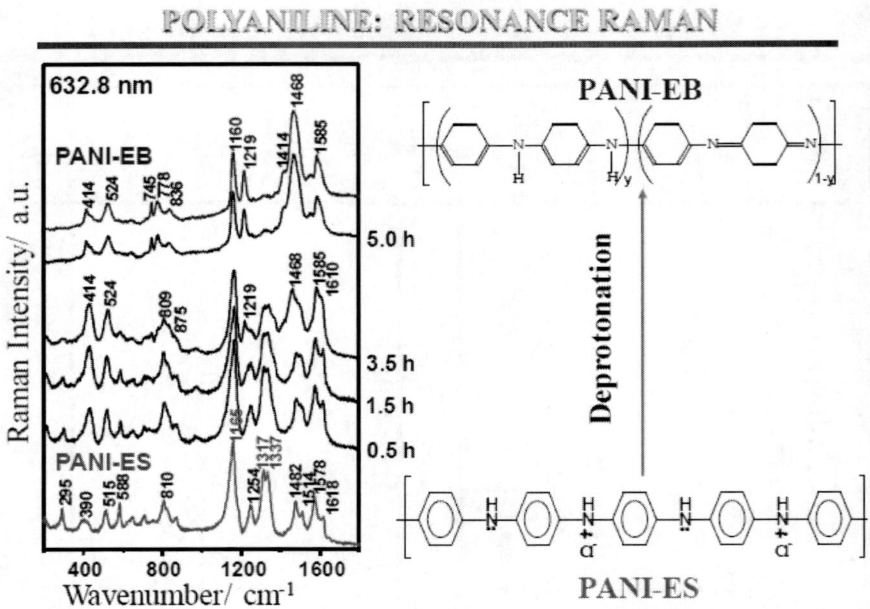

Figure 2.3. Raman spectra of PANI obtained during the deprotonation of PANI-ES at 632.8 nm laser line. Schematic representation of PANI structures before and after deprotonation are also displayed.

2.2. Nanostructured Polyanilines

The synthesis of nanostructured PANI, especially as nanofibers, can improve its electrical, thermal and mechanical stabilities. These materials can have an important impact for application in electronic devices and molecular sensors owing their extremely high surface area, synthetic versatility and low-cost. The conventional synthesis of polyaniline, based on the oxidative polymerization of aniline in the presence of a strong acid dopant, typically results in an irregular granular morphology that is accompanied by a very small percentage of nanoscale fibers [35, 36]. However, different approaches have been developed in order to produce PANI and many other polymers with nanostructured morphology. In this chapter will be analysed the synthetic routes that produce nanostructured PANI, mainly as nanofiber or nanotube morphology, without the use of rigid templates.

The nanostructured PANI has been prepared by different synthetic ways. Nevertheless, these approaches can be grouped into two general synthetic routes, as can be seen in the Figure 2.5.

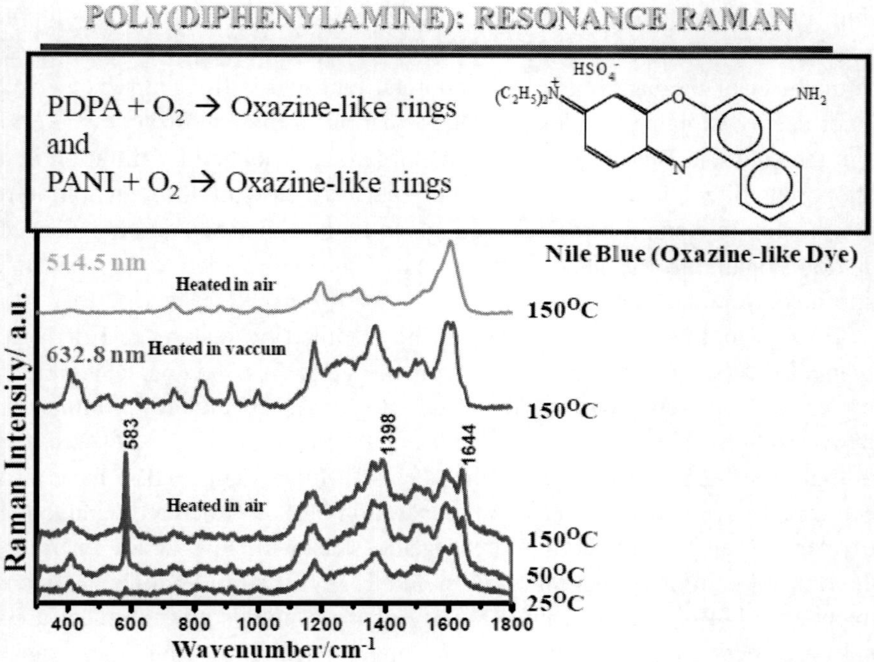

Figure 2.4. Resonance Raman spectra of PDPA-CSA at room temperature and heated at 50°C and 150°C in air and in vacuum obtained with exciting radiation 632.8 nm and 514.5 nm. [95] The chemical structure of Nile Blue, a typical dye with similar bands as observed for PANI-CSA, PDPA-CSA heated in air, is also given [34].

Uniform nanofibers of pure metallic PANI (30-120 nm diameter, depending on the dopant) have also been prepared by polymerization at an aqueous-organic interface. The first step (see item a) of the interfacial polymerization), the oxidant and monomers (aniline), dissolved in immiscible solvents, are put together without external agitation. Afterwards, some aniline monomers are oxidized in the interfacial region between the two solutions, being formed some oligomers (see item b) of the interfacial polymerization). It is hypothesized that migration of the product into the aqueous phase can suppress uncontrolled polymer growth by isolating the fibers from the excess of reagents. Afterwards, the initial chains grow up and more PANI chains are formed (see step c)). Interfacial polymerization can therefore be regarded as a non-template approach in which high local concentrations of both monomer and dopant anions at the liquid–liquid interface might be expected to promote the formation of monomer-anion (or oligomer-anion) aggregates. These aggregates can act as nucleation sites for polymerization, resulting in powders with fibrillar morphology. It has recently been demonstrated that the addition of certain surfactants to such an interfacial system grants further control over the diameter of the nanofibers. An important part that is frequently neglected or not deeply explained in details is the isolation of the nanostructured PANI from the solution. But, generally, the nanofibers are isolated by filtration in a nanoporous filters, being the isolated polymer washed with different solutions with the aim to clean it up. The solution can be also dialyzed and the cleaned solution containing the nanofibers is centrifugated in order to separate the nanofibers from the solution.

PANI nanofibers or nanotubes can be obtained by making use of large organic acids (see Figure 2.5). These acids form micelles upon which aniline is polymerized and doped (see Figure 2.5 steps (a), (b) and (c) of micellar polymerization). Fiber diameters are observed to be as low as 30-60 nm and are highly influenced by reagent ratios [37-40]. Ionic liquids (ILs) have also been used as synthetic media for the preparation of nanostructured conducting polymers [41-43]. Ionic liquids are organic salts with low lattice energies, which results in low melting points and many ILs are liquids at room temperature [44]. There is a large variety of ionic liquids and the most used ones are derived from imidazolium ring, pyridinium ring, quaternary ammonium and tertiary phosphonium cations. The usual differentiation between conventional molten salts and ionic liquids is based on the melting point. While most molten salts have melting points higher than 200°C, ionic liquids normally melt below 100°C [45].

The most unusual characteristic of these systems is that, although they are liquids, they present features similar to solids, such as structural organization at intermediate distances [46] and negligible vapor pressure [47]. This structural organization can act as a template like system, and PANI nanofibers are obtained when the aniline is polymerized in these media.

A broad variety of organic acids have been employed in order to modulate the diameter of PANI nanofibers (see Figure 2.5) [37-40]. The FTIR spectra of PANI doped with various organic acids, containing SO_3-H groups, show broad bands at about 3430 cm^{-1}, 1560 cm^{-1}, 1480 cm^{-1}, 1130 cm^{-1}, and 800 cm^{-1}, which are related to emeraldine PANI salt [48]. The UV-vis spectra of all doped PANI samples show two polaronic absorptions around 400 and 800 nm. The position of polaronic bands shifts to a long wavelength when the size of organic dopant increases. For instance, the polaron absorption for the PANI doped with smaller dopant (α-NSA, α-naphtalenesulfonic acid) is located at 800-900 nm. On the other hand, the polaron absorption for the doped PANI with larger dopant (β-NSA) is shifted to 1060-1118 nm.

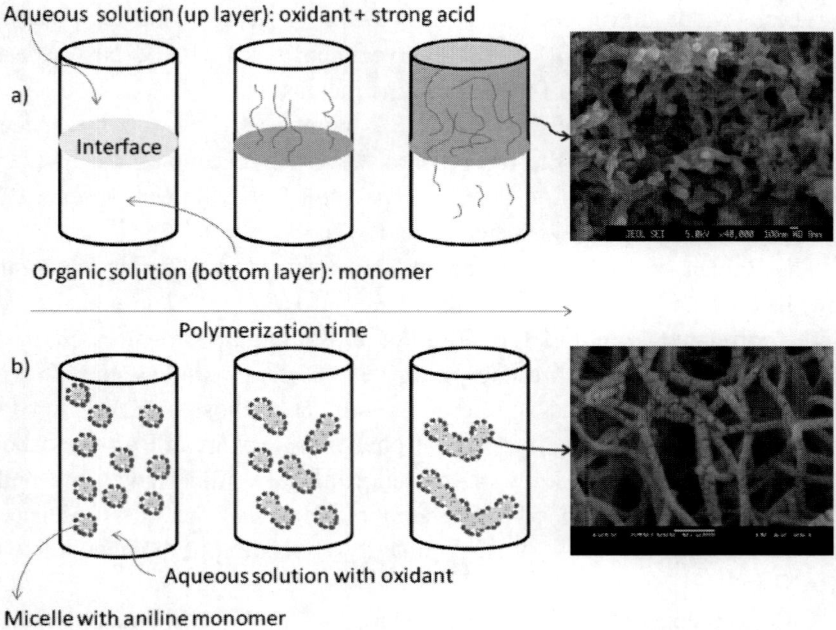

Figure 2.5. Schematic representation of: a) interfacial polymerization and b) micellar polymerization. In the interfacial polymerization the top layer is an aqueous solution containing HCl acid and $(NH_4)_2S_2O_8$ (others acids or oxidants can be used); the bottom layer has aniline dissolved in the chloroform (others solvents immiscible in water can be used). Starting the polymerization and migration of oligomers from organic bottom layer to the aqueous top layer and formation of PANI. The scanning electron microscopic (SEM) image was obtained from PANI nanofibers obtained from interfacial polymerization using HCl, $(NH_4)_2S_2O_8$, and chloroform. The nanofibers have ca. 30 nm of diameter. In the micellar polymerization the solubilization of aniline is in an aqueous solution containing organic acids that act as surfactants. After added the oxidant the polymerization starts and depending on the concentration of aniline in solution, it is possible to form hollow nanofibers (as named nanotubes) or nanofibers. The SEM image obtained from the PANI powder obtained from micellar polymerization using β-naphtalenesulfonic acid (β-NSA), $(NH_4)_2S_2O_8$, and molar ratio of β-NSA:aniline of 1:4. The nanofibers have ca. 93 nm of diameter.

The resonance Raman spectra for PANI-β-NSA nanofibers having different diameters show the same profile, it indicates that the morphological differences in PANI-NSA nanofibers have small influence in the Raman spectra from 1000 to 1800 cm^{-1}.

Comparing the RR spectra of PANI-NSA fibers to PANI-ES spectrum, bands at 1163 and 1330 cm^{-1} in PANI-NSA spectra can be associated with those at 1165 and 1317-1337 cm^{-1} in PANI-ES spectrum.

These bands have been assigned to βC-H and νC-N of polaronic segments, respectively [40]. Their relative intensities in PANI-NSA spectra increase as the molar ratio of β-NSA:aniline increases.

Hence, the RR data of the PANI-NSA nanofibers show that the spectral changes observed among the as-prepared PANI-NSA samples are owing to differences in the protonation degrees. The same behavior was observed for PANI nanofibers prepared with stearic acid [49].

The Raman spectra of PANI nanofibers prepared in micellar media also show the presence of bands at ca. 578, 1400, and 1632 cm^{-1}. These bands were strictly correlated with the formation of cross-linking structures in PANI chains after heating in the presence of air [34]. Different studies show that the bands at ca. 578, 1400, and 1632 cm^{-1} are similar to those observed for dyes with phenoxazine ring. The presence of phenoxazine rings in PANI backbone was also observed in the study of formation of polyaniline nanotubes under different acidic media [50, 51]. The authors concluded that the presence of phenoxazine units is crucial for stacking and stabilization of the nanotube wall of PANI [52].

On the comparison of the spectral behavior of PANI nanofibers/nanotubes prepared with NSA (β-naphtalenesulfonic acid) or with DBSA (dodecybenzenesulfonic acid) indicates that polymeric chains have a certain degree of extended conformation due to the presence of free-carrier absorption in the UV-VIS-NIR spectra. Hence, the presence of 609 cm^{-1} band in the PANI-NSA and PANI-DBSA Raman spectra indicates that these samples have a certain degree of extended conformation. The band at 609 cm^{-1} can be assigned to a vibrational mode related to benzene deformations or torsions. Probably, this mode is sensible to changes of the dihedral angle between neighbors benzene rings, or in other words, sensible to the conformation of the PANI chains [52].

Electron microscopic images reveal the loss of the fibrous morphology of PANI after treatment of PANI-NSA samples with HCl solution in order to acquire higher doping state [40]. However, further studies reveal that submitting the PANI-NSA to heating treatment at 200°C, occurs the formation of a high degree of cross-linking structures, verified by the appearance of characteristic RR bands at 578, 1398 and 1644 cm^{-1}, hence the fibrous morphology is retained after the doping process [40, 53]. PANI nanofibers synthesized in ionic liquids have been studied by Raman spectroscopy. PANI nanofibers were obtained by electropolymerization of aniline in $BMIPF_6$ (1-butyl-3-methyl-imidazolium hexafluorophosphate) [54]. The Raman spectra show that the PANI is similar to the emeraldine salt form.

However, the intensity of the quinoid ring stretching at 1578 cm^{-1} is higher than that of the benzenoid band at 1469 cm^{-1}, indicating the existence of a higher amount of quinoid structures. The authors suggest that the PANI film synthesized in this ionic liquid media is formed by small amount of non-conducting forms such as PANI-EB and PANI-PB [54].

PANI nanofibers prepared from interfacial polymerization were also characterized by Raman spectroscopy. It was observed that the bands at 200 and 296 cm^{-1}, related to C_{ring}-N-C_{ring} deformation and lattice modes of polaron segments of PANI with type-I crystalline arrangement [55], practically disappears in the Raman spectra of PANI nanofibers. This effect is very pronounced for the nanofiber sample prepared using 5.0 $mol.L^{-1}$ HCl aqueous solution. The bands at about 400 cm^{-1} indicates the increase of the torsion angles of the C_{ring}-N-C_{ring} segments (see Figure 2.6).

The FTIR spectra for PANI nanofibers display higher changes in the region from 2000 to 4000 cm^{-1} [55]. Mainly the bands related to NH_2^+ modes at 2480, 2830, and 2920 cm^{-1} increase in their intensities for PANI samples prepared with higher HCl concentration (higher than 1.0 $mol.L^{-1}$), consequence of the increase of protonated imine and amine nitrogens in the structure of PANI. The band at 3200 and 3450 cm^{-1}, also change their relative intensities, can be assigned to bonded N-H and free N-H stretching modes [56, 57]. The changes in the IR bands associated with an increase in the torsion angles of C_{ring}-N-C_{ring} segments is owing to the formation of bipolarons (protonated, spinless units) in the PANI backbone higher than the PANI samples prepared by the conventional route. The nanostructured surface of PANI permits major diffusion of the ions inside the polymeric matrix leading to a more effective protonation of the polymeric chain than the PANI prepared in the conventional way, leading to the reduction of crystallinity of PANI, and the decrease in the amount of nanofibers [55].

CONCLUSION

The screening of the electronic and vibrational structure of the polyaniline nanofibers has been decisive in the studies related to the formation, interactions between the chains, properties and stabilities of the nanostructured polyaniline. Nowadays, two great approaches are used to acquire the PANI with nanostructured morphology without the use of rigid hosts: (i) polymerization of aniline in a micellar media and (ii) polymerization of aniline on the interface between two solvents.

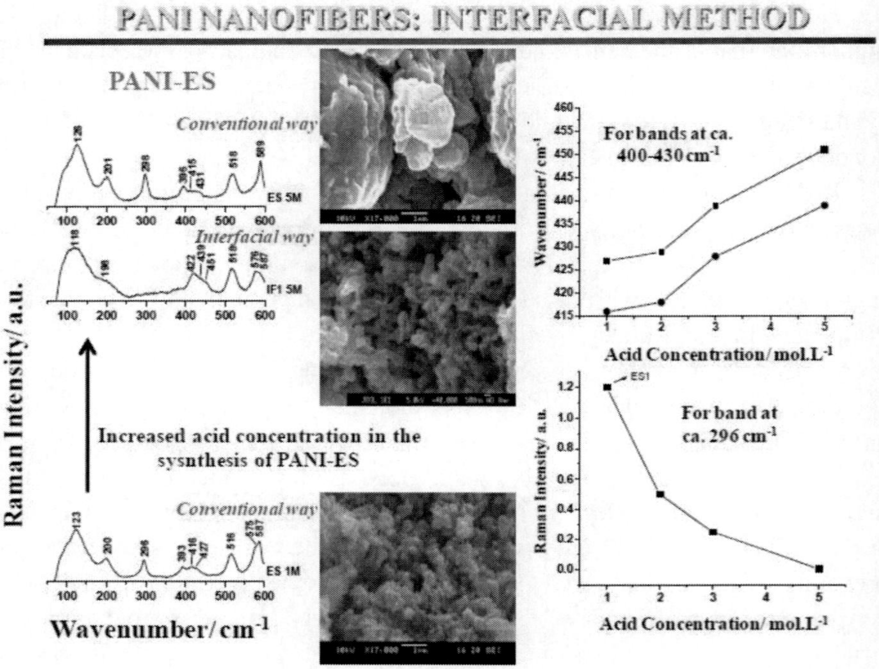

Figure 2.6. Resonance Raman spectra of PANI prepared from conventional and interfacial methods obtained at 632.8 nm laser line. The SEM images of each sample are also displayed. For comparison purposes the changing of band intensity at ca. 296 cm^{-1} and the shift of bands at ca. 400-430 cm^{-1} are plotted as a function of the HCL acid concentration used in the PANI synthesis.

However, the morphology of PANI obtained without rigid hosts is more susceptible to the synthetic conditions (such as pH) and also post-synthesis procedures. Mainly, it is observed shifts in the vibrational frequencies of polyaniline and also variations in their intensities. The presence of bands owed to phenoxazine rings is observed in PANI backbone formed in micellar media. The presence of phenoxazine units is crucial for stacking and stabilization of the nanotube wall of PANI. Probably, The π-π stacking formed by phenoxazine rings, in the PANI backbone prepared in micellar media, is one of the driving forces for the formation of PANI chains with extended conformation and PANI particles with one-dimensional (needles and/or nanofibers) morphology. The changes in the intensities of the vibrational spectra at low energies are associated with an increase in the torsion angles of C_{ring}-N-C_{ring} segments due to the formation of bipolarons (protonated, spinless units) in the PANI backbone higher than the PANI samples prepared by the conventional route. The nanostructured surface of PANI permits major diffusion of the ions inside the polymeric matrix leading to a more effective protonation of the polymeric chain than the PANI prepared in the conventional way, leading to the reduction of crystallinity of PANI, and the decrease in the amount of nanofiber.

REFERENCES

[1] Shirakawa, H., Ikeda, S. *Polymer J.* 1971, 2, 231.

[2] Shirakawa, H., Ikeda, S. *J. Poylm. Sci. Chem.* 1974, 12, 929.

[3] Chiang, C. K., Druy, M. A., Gau, S. C., Heeger, A. J., Louis, E. J., MacDiarmid, A. G., Park, Y. W., Shirakawa, H. *J. Am. Chem. Soc.* 1978, 100, 1013.

[4] Chiang, C. K., Fincher Jr, C. R., Park, Y. W., Heeger, A. J., Shirakawa, H., Louis, E. J., MacDiarmid, A. G. *Phys. Rev. Lett.* 1977, 39, 1098.

[5] Shirakawa, H. *Angew. Chem. Int. Ed.* 2001, 40, 2575.

[6] Shirakawa, H., Louis, E. J., MacDiarmid, A. G., Chiang, C. K., Heeger, A. J. *J. Chem. Soc.: Chem. Commun.* 1977, 16, 578.

[7] MacDiarmid, A. G. *Angew. Chem. Int. Ed.* 2001, 40, 2581.

[8] Nigrey, P. J., MacDiarmid, A. G., Heeger, A. J. *J. Chem. Soc.: Chem. Commun.* 1979, 14, 594.

[9] Han, C. C., Elsenbaumer, R. L. *Synth. Met.* 1989, 30, 1, 123.

[10] Heeger, A. J. *Angew. Chem. Int. Ed.* 2001, 40, 2591.

[11] MacDiarmid, A. G., Epstein, A. J. *Faraday Discuss. Chem. Soc.* 1989, 88, 317.

[12] MacDiarmid, A. G., Epstein, A. J., In: *Conducting polymers, emerging technologies,* Technical Insights: New Jersey, 1989, page 27.

[13] MacDiarmid, A. G., Chiang, J. C., Richter, A. F., Sonosiri, N. L. D., *Conducting Polymers*, Alcácer, L. Ed., Reidel Publications: Dordrecht, 1989.

[14] MacDiarmid, A. G., Epstein, A. J. In: *Frontiers of Polymers and Advanced Materials,* Prasad, P. N. Ed., Plenum Press: New York, 1994, 251.

[15] Huang, W. S., MacDiarmid, A. G. *Polymer* 1993, 34, 1833.

[16] Batchelder, D. N. In: *Optical Techniques to Characterize Polymer Systems;* H. Brässler, Eds.; Elsevier: Amsterdam, 1987.

[17] Batchelder, D. N., Bloor, D. In: *Advances in Infrared and Raman Spectroscopy*, Wiley-Heyden: London, 1984.

[18] Clark, J. H., Dines, T. J., *Angew. Chem. Int. Ed. Engl.* 1986, 25, 131.
[19] McHale, J. L., *Molecular Spectroscopy,* Prentice-Hall: US, 1999.
[20] Sariciftci, N. S., Bartonek, M., Kuzmany, H., Neugebauer, H., Neckel, A., *Synth. Met.* 1989, 29, 193.
[21] Furukawa, Y., Ueda, F., Hydo, Y., Harada, I., Nakajima, T., Kawagoe, T., *Macromolecules* 1988, 21, 1297.
[22] Quillard, S., Louarn, G., Lefrant, S., MacDiarmid, A. G. *Phys. Rev. B* 1994, 50, 12496.
[23] Berrada, K., Quillard, S., Louarn, G., Lefrant, S. *Synth. Met.* 1995, 69, 201.
[24] Louarn, G., Lapkowski, M., Quillard, S., Pron, A., Buisson, J. P., Lefrant, S. *J. Phys. Chem.* 1996, 100, 6998.
[25] McCall, R. P., Ginder, J. M., Leng, J. M., Ye, H. J., Manohar, S. K., Masters, J. G., Asturias, G. E., MacDiarmid, A. G., Epstein, A. J. *Phys. Rev. B* 1990, 41, 5202.
[26] Pereira da Silva, J. E., Córdoba de Torresi, S. I., De Faria, D. L. A., Temperini, M. L. A. *Synth. Met.* 1999, 101, 834.
[27] Pereira da Silva, J. E., De Faria, D. L. A., Córdoba de Torresi, S. I., Temperini, M. L. A. *Macromolecules* 2000, 33, 3077.
[28] Cochet, M., Louarn, G., Quillard, S., Boyer, M. I., Buisson, J. P., Lefrant, S., *J. Raman Spectrosc.* 2000, 31, 1029.
[29] Cochet, M., Louarn, G., Quillard, S., Buisson, J. P., Lefrant, S., *J. Raman Spectrosc.* 2000, 31, 1041.
[30] Quillard, S., Berrada, K., Louarn, G., Lefrant, S., Lapkowski, M., Pron, A. *New J. Chem.* 1995, 19, 365.
[31] Engert, C., Umapathy, S., Kiefer, W., Hamaguchi, H. *Chem. Phys. Lett.* 1994, 218, 87.
[32] Niaura, G., Mazeikiene, R., Malinauskas, A. *Synth. Met.* 2004, 145, 105.
[33] Do Nascimento, G. M., Temperini, M. L. A. *J. Raman Spectrosc.* 2008, 39, 772.
[34] Do Nascimento, G. M., Pereira da Silva, J. E., Córdoba de Torresi, S. I., Temperini, M. L. A. *Macromolecules* 2002, 35, 121.
[35] Huang, J., Kaner, R. B. *Angew. Chem. Int. Ed.* 2004, 43, 5817.
[36] Huang, J., Kaner, R. B. *J. Am. Chem. Soc.* 2004, 126, 851.
[37] Zhang, Z. M., Wei, Z. X., Wan, M. X. *Macromolecules* 2002, 35, 5937.
[38] Qiu, H. J., Wan, M. X., Matthews, B., Dai, L. M. *Macromolecules* 2001, 34, 675.
[39] Wei, Z. X., Wan, M. X. *Adv. Mater.* 2002, 14, 1314.

[40] Do Nascimento, G. M., Silva, C. H. B., Temperini, M. L. A. *Macromol. Rapid Commun.* 2006, 27, 255.

[41] Gao, H., Jiang, T., Han, B., Wang, Y., Du, J., Liu, Z., Zhang, J. *Polymer* 2004, 45, 3017.

[42] Rodrigues, F., Do Nascimento, G. M., Santos, P. S. *Macromol. Rapid Commun.* 2007, 28, 666.

[43] Rodrigues, F., Do Nascimento, G. M., Santos, P. S. *J. Electron Spectrosc. Rel. Phenom.* 2007, 155, 148.

[44] Davis Jr., J. H., Gordon, C. M., Hilgers, C., Wasserscheld, P. In: *Ionic Liquids in Synthesis*, Wassercheid, P., Welton, T., Eds.; Wiley-VCH: New York, 2002, page 7.

[45] Wasserscheid, P., Keim, W. *Angew. Chem. Int. Ed.* 2000, 39, 3773.

[46] Dupont, J. *J. Braz. Chem. Soc.* 2004, 15, 3, 341.

[47] Earle, M. J., Esperança, J. M. S. S., Gilea, M. A., Lopes, J. N. C., Rebelo, L. P. N., Magee, J. W., Seddon, K. R., Wildegren, J. A. *Nature* 2006, 439, 831.

[48] Huang, J., Wan, M. X. *J. Polym. Science part A: Polym. Chem.* 1999, 37, 1277.

[49] Wang, X., Liu, J., Huang, X., Men, L., Guo, M., Sun, D. *Polym. Bull.* 2008, 60, 1.

[50] Trchova, M., Syedenkova, I., Konyushenko, E. N., Stejskal, J., Holler, P., Ciric-Marjanovic, G. *J. Phys. Chem. B* 2006, 110, 9461.

[51] Stejskal, J., Sapurina, I., Trchova, M., Konyushenko, E. M., Holler, P. *Polymer* 2006, 47, 8253.

[52] Do Nascimento G. M., Silva, C. H. B., Izumi, C. M. S., Temperini, M. L. A. *Spectrochim. Acta Part A* 2008, 71, 869.

[53] Do Nascimento, G. M., Silva, C. H. B., Temperini, M. L. A. *Polym. Degrad. Stab.* 2008, 93, 291.

[54] Wei, D., Kvarnstrom, C., Lindfors, T., Ivaska, A. *Electrochem. Commun.* 2006, 8, 1563.

[55] Do Nascimento, G. M., Kobata, P. Y. G., Temperini, M. L. A. *J. Phys. Chem. B* 2008, 112, 11551.

[56] Jana, T., Roy, S., Nandi, A. K. *Synth. Met.* 2003, 132, 257.

[57] Lunzy, W., Banka, E. *Macromolecules* 2000, 33, 425.

In: Advances in Conducting Polymers Research ISBN: 978-1-63463-258-4
Editor: Laura Michaelson

Chapter 2

CONDUCTING POLYMER MICRO-/ NANO- STRUCTURES VIA TEMPLATE-FREE METHOD

Hang-Jun Ding[1,*], Yun-Ze Long[2,4,†], Zhi-Ming Zhang[3], Huai Yang[5], Gui-Feng Yu[2,6] and Zhou Yang[1]

[1]School of Materials Science & Engineering,
University of Science & Technology Beijing, Beijing, P. R. China
[2]College of Physics, Qingdao University, Qingdao, P. R. China
[3]College of Chemistry and Chemical Engineering,
Ocean University of China, Qingdao, P. R. China
[4]State Key Laboratory Cultivation Base of New Fiber Materials and
Modern Textile, Qingdao University, Qingdao, P. R. China
[5]School of Engineering, Peking University, Beijing, P. R. China
[6]College of Science and Information, Qingdao Agricultural University,
Qingdao, P. R. China

† E-mail address: yunze.long@163.com or yunze.long@qdu.edu.cn.
* E-mail address: dinghangjun@163.com.

ABSTRACT

This chapter briefly summarizes recent advances in synthesis, characterization of conducting polyaniline (PANI) micro-/nanostructures (e.g., hollow tubes and spheres) via template-free method. The synthesis strategies, self-assembly mechanism and process parameters for the template-free method are discussed. It is found that the morphology (tubes, wires/fibers, and spheres) and size of the PANI micro-/nanostructures can be controlled by adjusting experimental parameters. For example, PANI nanofibers with 10 nm in average diameter have been successfully fabricated. In particular, superhydrophobic films (contact angle can reach up to 148.0^{o}) composed of mono-dispersed PANI nanospheres and oriented-arrays of PANI nanospheres with high crystallinity have also been prepared by this simple and versatile template-free method.

1. INTRODUCTION

Conducting polymers (CPs) such as polyaniline (PANI), polypyrrole (PPY), poly(3,4-ethylene dioxythiophene) (PEDOT), and poly(3-hexylthiophene) (P3HT), have been recognized as promising organic semiconductors due to their controllable chemical/electrochemical properties, light weight, low cost, good biocompatibility, facile processability, and adjustable electrical conductivities from 10^{-11} to 10^{5} S cm^{-1} upon doping, reversible doping/dedoping process. Particularly, CP micro-/nanostructures such as wires, tubes, fibers and spheres have drawn much attention due to their interesting nanosize-related properties (e.g., larger specific surface area) and useful applications (e.g., electronic and biomedical materials, protective clothing, filtration media, charge storage devices and sensors and actuators). CP micro-/nanostructures can be prepared by a variety of methods such as hard physical template-guided synthesis, interfacial polymerization, dilute polymerization, reverse emulsion polymerization, template-free method, nanoprinting, electrospinning, *etc*. Here, we briefly introduce the discovery of template-free method. In 1998, Prof. Meixiang Wan accidentally discovered that PANI nanotubes could be prepared by conventional *in-situ* polymerization in the presence of β-naphthalene sulfonic acid (β-NSA) as the dopant without using any membrane as template [1]. In this method, the conducting polymer monomer (aniline, pyrrole, EDOT) and dopant were firstly dissolved in the solvent, and oxidant (e.g. APS, $FeCl_3$) was added into the mixture. Conducting polymer micro-/nanostructures could be obtained after a self-assembly process. Due to omitting membrane as template, this method was latterly called as “template-free method”. Compared with hard-template method [2], the template-free method is simple and inexpensive because of omitting template and post-treatment of removing template. They systematically studied synthesis method, structural characteristics, physical properties and potential applications by this method and they demonstrated the universality of this method for PANI and PPY micro-/nanostructures by changing polymer chain, polymerization method (chemical and electrochemical polymerization) and dopant nature. They found not only aniline monomer, aniline derivatives nanotubes could also be synthesized by template-free method. For example, they used template-free method to synthesize ortho-toluidine (OT) instead of aniline with β-NSA as dopant and the POT-β-NSA microtubes with 0.8 and 6.0 μm in diameter could also be obtained [3]. Not only electrochemical template method [4], but also chemical template-free method [5] could be used successfully to synthesize PPY-NSA microtubes. Many kinds of morphologies, including nanotubes, nanofibers, microspheres, were later synthesized with different dopants [6,7], indicating the template-free method is a reliable and practical method of synthesis tubular PANI and their dereivatives.

2. Self-Assembly Mechanism of the Template-Free Method

It is known when template-free method is employed to prepare nanostructures of the CPs, formation and growth of the nanostructures is a self-assembly process because of omitting hard-template. Molecular interactions, such as hydrogen bonds, Van der Waals forces, stacked interaction, are usually served as powerful driving forces for self-assembly of the nanostructures in the absence of hard-templates [8].

2.1. Self-assembly Mechanism of the Normal Template-free Method

In general, surfactant is a common “soft-template” because it is easy to form thermodynamically stable and controllable nanoscale dimensions in solution or at interface. The reagents in template-free method include conducting monomer, dopant and oxidant. The micelles in aqueous solution can be formed by dopant, dopant/monomer salt or super-molecule and monomer itself due to hydrophilic dopant (e.g., $-SO_3H$ group) and dopant/monomer salt or amphiphilic molecule of monomer (e.g. aniline). Prof. Wan and her cooperators studied the formation process of PANI-β-NSA nanotubes via template-free method conducting polymer (PANI or PPY) and found the micelles formed by dopant act as "soft-templates" in formation of conducting polymer nanostructures due to hydrophilic group of $-SO_3H$ and hydrophobic group $-C_{10}H_7$ of NSA. dopant/monomer salt or super-molecule and monomer itself due to hydrophilic dopant (e.g., $-SO_3H$ group) and dopant/monomer salt or amphiphilic molecule of monomer (e.g. aniline). They provided three positive evidences of existence of the micelles in the formation of soft-templates [9].

(1) They used dynamic light scattering (DLS) to demonstrate the existence of micelles in the solution and the diameter of both nanotubes and micelles were found to increase with increase of the [NSA]/[An] ratios [9], indicating that the micelle act as templates for the nanotubes.
(2) In general, the micelle size can be adjusted by changing the ionic strength of solution or the polarity of the solvent [10]. The existence of the micelles was proven by the fact that the diameter of the nanotubes increased when KCl aqueous solution was added as solvent.
(3) Moreover, the cylindrical shape of the micelles were directly measured by freeze-fracture transmission electron microscope (TEM) [11], which provide the direct proofs of the existence of micelles in the solution. The micelles could also be formed by dopant/monomer salt in the solution [12]. As we know, dopant/monomer salt is easily formed in the reaction through an acid/base reaction due to basic monomer and acidic dopant. All shape peaks observed from XRD of the aniline-NSA salt could be found in the PANI-NSA microtubes [13], proving the aniline salt exists in the PANI microstructures. Spherical micelles composed of anilinium cations may be formed as "soft-template" [14] in the formation PANI nanotubes in the presence of inorganic acids as dopant [15]. Therefore, micelle model [16] can interpret the formation of various micro/nanostructures of the conducting polymers prepared by template-free method. Therefore,

the micelles composed by dopant, dopant/monomer, and aniline monomer, can act as “soft-templates” in the formation of conducting polymer micro-/nanostructures. Spherical micelles are firstly formed in the initial stage due to the low surface energy and aggregate to form cylindrical or flat bilayer [17]. Monomer can be diffused into micelles and form monomer filled micelles. Once oxidant is added, polymerization only takes place at water/micelle interface because of hydrophilic oxidant. Growth of the micro/nanostructures is controlled by accretion and elongation process [18]. Competition between the micelles and molecular interaction (e.g. hydrogen bonds, staked and hydrophobic interactions) will result in various micro/nanostructures, including spheres, nanotubes, nanofibers or nanofiber junctions.

2.2. Self-assembly Mechanism of the Simplified Template-Free Method (STFM)

PANI has a special proton doping mechanism that results in formation of delocalized poly-semiquinone radical cations and accompanied with enhancement of conductivity by 10^{10}. [19] Therefore acidic dopant is generally used in the reaction solution in order to form conductive PANI (i.e. the emeraldine salt form). As mentioned above, aniline may exist in the form of anilinium cations or free aniline in the reaction solution as micelles. When ammonium peroxydisulfate ($(NH_4)_2S_2O_8$, APS) is used as the oxidant the pH value of the reaction solution decreases with increase of the polymerization time, suggesting that the proton is produced during polymerization due to reaction of APS with aniline monomer. Based on this, author continued to simplify the template-free method to synthesize PANI nanotubes (as shown in Figure 1) with conductivity of 30.6 S/cm only using APS as oxidant without using template and adding acidic dopant. This approach is called as simplified template-free method (STFM). To our best knowledge, this is the simplest approach to prepare PANI nanotubes because of being not only omitting template, but also simplifying reaction regents.

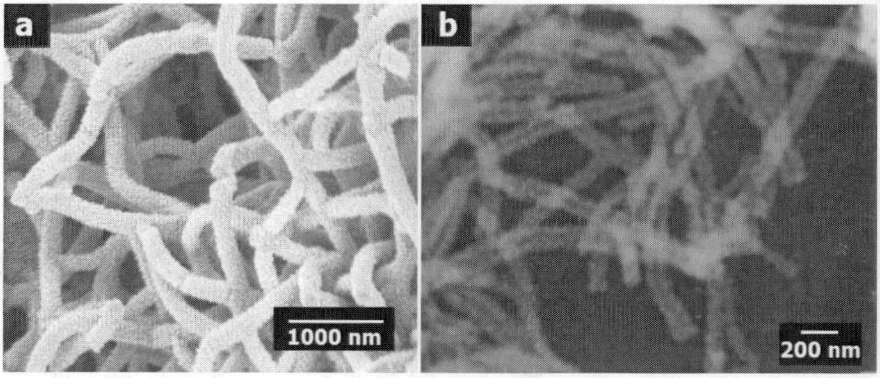

Figure 1. Typical (a) SEM and (b) TEM images of the PANI nanotubes prepared by simplified template-free method (STFM).

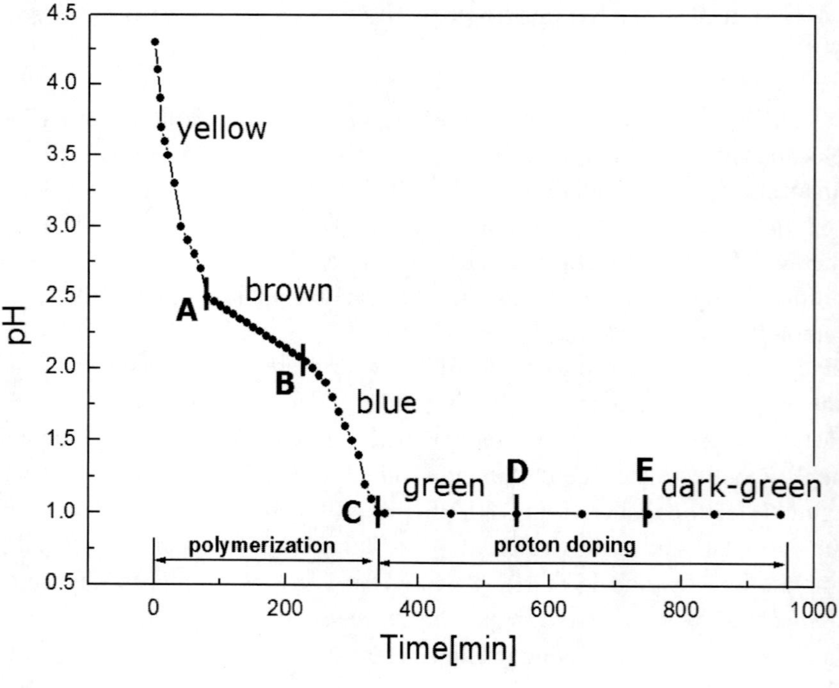

Figure 2. Variation of the pH value with the polymerization time and accompanied by the color change of PANI synthesized by STFM. Other reaction conditions: [An]= 0.06M, [APS]/[An] = 1:1.

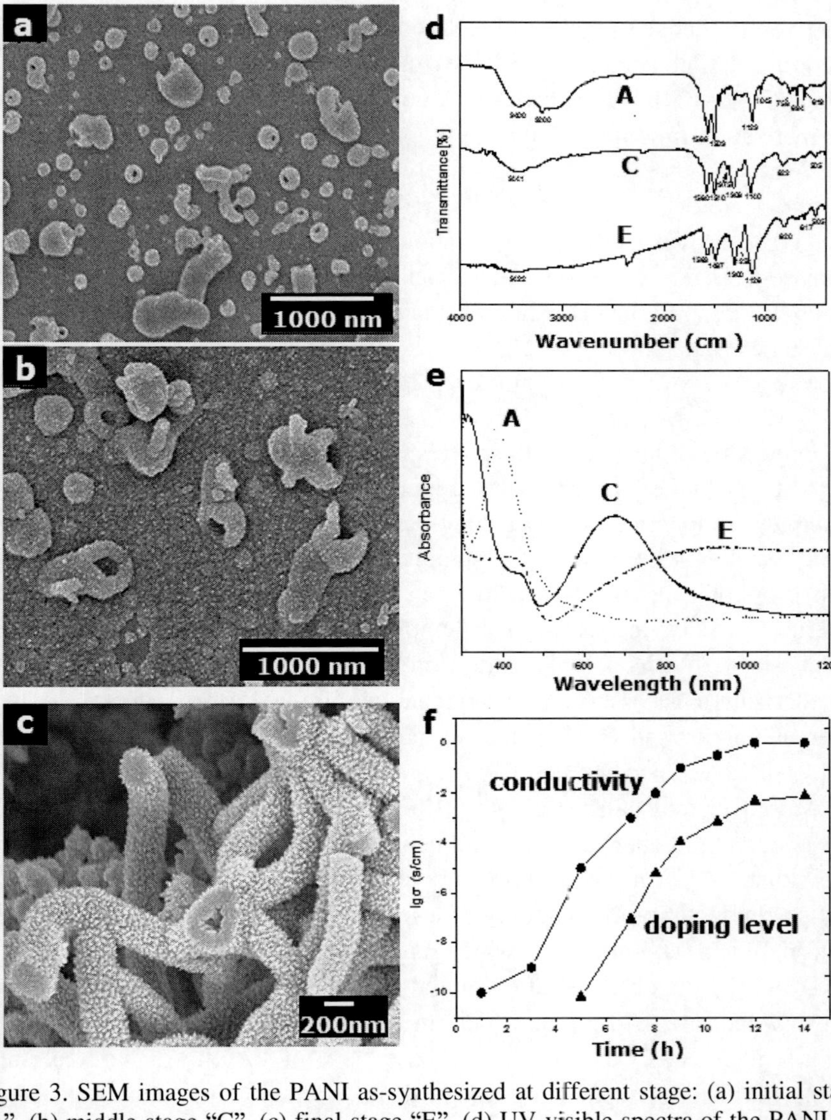

Figure 3. SEM images of the PANI as-synthesized at different stage: (a) initial stage "A". (b) middle-stage "C". (c) final-stage "E". (d) UV-visible spectra of the PANI as-synthesized at the "A", "C", "E" stage; (e) FTIR spectra of the PANI as-synthesized at different stage of "A", "C", and "E" stage; (f) Dependence of the conductivity and the doping level for PANI nanotubes synthesized by STFM.

In order to understand formation mechanism of the nanotubes prepared by STFM, polymerization process with time was measured. It is found that the color and pH value of the reaction solution of the as-synthesized PANI are changed with polymerization time. For instance the color of the solutions underwent a change of yellow-blue-green. Simultaneously, the pH value of the reaction solution decreases with polymerization time due to production of sulfuric acid [20]. As shown in Figure 2, the curve of pH value with polymerization time divides into three stages of initial (“A”), middle (“B-C”) and final (“C-D-E”) state, respectively. The pH value at the initial stage sharply decreased from 4.3 to 2.5 and the color of the reaction solution changed to brown. With polymerization process, the pH value continued to decrease from 2.5 to 1.0 and color changed from blown to blue at the middle stage (“B-C”) until the pH value remained at 1.0 and color became green at the final stage (“D-E”).

Morphology of the as-synthesized PANI at initial, middle and final stage was measured. As shown in Figure 3, open-spheres at the “A” stage (Figure 3a), short tubes composed with spheres linearly aggregated at the “C” (Figure 3b) and long tubes at “E” stage (Figure 3c) are observed, respectively. In principle, the spherical micelles composed of aniline monomer can be formed in the reaction solution due to amphiphilic molecule of aniline resulted from its hydrophobic group of benzene ring and hydrophilic -NH_2 group. Those spherical micelles might act as soft-templates in the formation of spheres at the initial stage. With polymerization process, the proton produced during polymerization accesses linear aggregation of the formed spheres to form tubes at middle and final stages via polymerization and elongation process [21].

Moreover molecular structure of the nanotubes with polymerization time also underwent a change from oligomer to emeraldine base form (EB) and to emeraldine salt form (ES) of PANI proved by UV-Visible and FTIR spectra as well as conductivity measured by four-probe method. As shown in Figure 3d, a strong band at ~400 nm, which is assigned as p-amino-diphenylamine (PADPA) [22], is observed at the initial stage (“A”). Except for a band at 420 nm, two bands of 340 nm and 650 nm at the middle stage “C” are observed that are identical to the emeraldine base form of PANI. At the final stage (“E”), the band at 650 nm assigned as excitation of quinoid segment disappeared, while a band at >800 nm with a long tail regarded as polaron band is observed that is identical to the emeraldine salt form [23]. FTIR spectra also show that character bands at 696 and 754 cm^{-1} characteristic of mono-substituted aromatic rings, which is assigned as oligomer of aniline due to only presence in the terminal of the short-chain oligomer [24], are observed from product as-synthesized at “A” stage and their intensity decreases with the increase of the chain length. Moreover, small and weak peaks in the region of 1600 cm^{-1} and 500 cm^{-1} assigned as oligomer-aniline [25] are also observed in

"A' stage expect for two major stronger adsorption at about 1570 cm^{-1} and 1500 cm^{-1}, attributed to the quinoid ring and benzenoid ring stretching respectively (Figure 3e). Those results suggest that the nanostructures as-synthesized at "A" or "B" stage might be regarded as an oligomer of PANI. At the middle-stage "C", there is a unique weak C-N stretching adsorption at about 1375 cm^{-1}, indicating consisting of about equal amounts of the quinoid and benzenoid units [26]. Furthermore, the intensity of the band at 1140 cm^{-1}, which is assigned as the doped PANI [27], is weaker, suggesting that the middle-product "C" is the emeraldine base (EB) from of PANI. At the "E" stage, the FTIR spectra are identical to the emeraldine salt form of PANI, especially the strength of the band at 1141 cm^{-1} gradually increased from "C" to "E", indicating that the doping level increase with polymerization. Above-described variations of the molecular structure with the polymerization time are also proved by a fact that the conductivity and corresponding doping degree measured by XPS are enhanced by increasing polymerization time as shown in Figure 3f. Above-mentioned results further indicate that the polymerization process mainly takes place at the initial ('A") and middle stage ("B-C"), while the proton doping process is carried out at the final stage ("C-D-E"). It has been demonstrated that H_2SO_4 produced during the polymerization provide large amount of protons to dope PANI, resulting in formation of the emeraldine salt form with a high conductivity.

3. Controlling the Morphology and Diameter

As we know, controlling the morphology and size of the micro-/nanostructures synthesized by template-free method is very difficult because of omitting hard membrane as templates. However, the morphology and size of the micelles are strongly affected by polymerization condition, such as concentration of dopant, monomer and oxidant, the molar ratio of dopant and oxidant to monomer, and temperature and so on. Therefore, it is reasonable to believe that morphology and diameter of the template-free synthesized micro/nanostructures could be controlled by polymerization condition.

3.1. Controlling the Morphology of CP Nanostructure

It is found that the morphology, such as spheres, tubes and junctions, of CP nanostructures via template-free method could be controlled by changing the polymerization condition. For example, PANI-P-NSA could change from hollow spheres to nanotubes when the temperature increased from -10°C to room temperature [28]. The PANI-SA hollow spheres could change to nanotubes as the molar ratio of SA to aniline decreased from 1.0 to 0.1 [29]. Based on micelles model, the micelles formed by aniline/SA are served as soft-templates in formation of either naotubes or hollow spheres of the PANI-SA through a self-assembly process. However, the hydrogen bond of the –OH group of SA with the amine of PANI might be a driving force results in change of the morphology from hollow spheres to nanotubes as increase of the molar ratio of SA to aniline from 1.0 to 0.1 [30].

3.2. Controlling the Size of CP Nanostructure

Nowadays one-dimensional (1D) nanostructured conducting polymers have become the focus of intensive research owing to their highly conjugated molecular structure, metal-like conductivity and promising application in nanomaterials and nanodevices. As we all known, the nano-scaled materials have a unique size effect, which can strongly affect their chemical and physical properties. Thus preparation or synthesis for controllable diameter and morphology of the nanostructured conducting polymers is a challenge for material scientists. "hard-template" method is one effective and straightforward route to fabricate controllable diameter of conducting polymer nanostructures by the porous diameter of the template, however, the preparation and removal of the template is rather tedious. Although synthesis of well-controlling nanostructures by template-free method is difficult due to omitting membrane as templates, it is found that the diameter of the micro/nanostructures is affected by polymer and dopant structure as well as polymerization conditions, particularly the molar ratio of monomer to dopant and the concentration of dopant [31]. It is found that the average diameter of PANI-β-NSA nanotubes increases with the increase in dopant concentration when the dopant concentration is larger than 2.5×10^{-2} mol/L. At the same concentration, moreover, the average diameter of the nanotubes followed an order of PANI-α-NSA< PANI-β-NSA< PANI-(1,5-NSA) [32]. Dopant structure and size will play an important role in controlling the diameter of the template-free synthesized nanostructures. Four fatty acids with different number of $-CH_2$ group were used as dopants to understand their influence on morphology and diameter of the PANI nanostructures by template-free method [33] and it is found that the diameter of the nanofibers is controllable, showing the diameter increases with increase of the number of $-CH_2$ group of alkyl chain [33]. PANI nanostructures can also be controlled by polymerization condition, particularly changing the molar ratio of dopant to aniline [34].

3.3. Quantitatively Controlling the Size of CP Nanofibers

Although these above results suggest that the diameter of the PANI nanofibers by template-free method can be adjusted by the polymerization parameters, simple and quantitative approach to control the size of conducting polymer nanostructures via template-free method is necessary.

As we know, the emeraldine salt form of PANI (i.e. its conducting state) is prepared by oxidative polymerization in a strong acidic medium [35]. Thus influence of the oxidant on structure and chemical or physical properties of PANI is expected. Various oxidants, including ammonium peroxydisulfate ($(NH_4)_2S_2O_8$, APS) [36], tetrabutylammonium persulfate (TBAP) [37], hydrogen peroxide (H_2O_2) [38], benzoyl peroxide [39], ferric chloride ($FeCl_3$) [40], and chloroaurate acid ($HAuCl_4$) [41] have been used to synthesize PANI, previous reported results mainly emphasized the effect of oxidants on the polymerization yield, and APS was regarded as the optimal oxidant for PANI [42] due to its high yield.

It is found that the diameter of the template-free synthesized nanofibers using APS is in the range of 150-200 nm. However, when ferric chloride ($FeCl_3$) was used as oxidant, the diameter of the self-assembly PANI nanofibers was only in the range of 17-20 nm [43], which is reduced by 10 times compared with that of nanotubes oxidized by APS. According to micelles model, it is reasonable to accept that micelles composed of dopant and/or aniline monomer formed in the solution are served as the soft-templates in the formation of nnaofibers. Since the redox potential of $FeCl_3$ (0.77) is lower than that of APS (2.0), the oxidized polymerization rate by $FeCl_3$ might be much slower compared with that of APS, resulting in the much smaller diameter of the nanofibers oxidized by $FeCl_3$ than that of nanofibers oxidized by APS. Therefore, using the oxidants with different oxidant potential might be an efficient way to control the diameter of the conducting polymer nanostructures by template-free method. Herein, different oxidants, APS, $Ce(SO_4)_2$, $FeCl_3$, $Fe_2(SO_4)_3$, and $CuCl_2$ were used as oxidants.

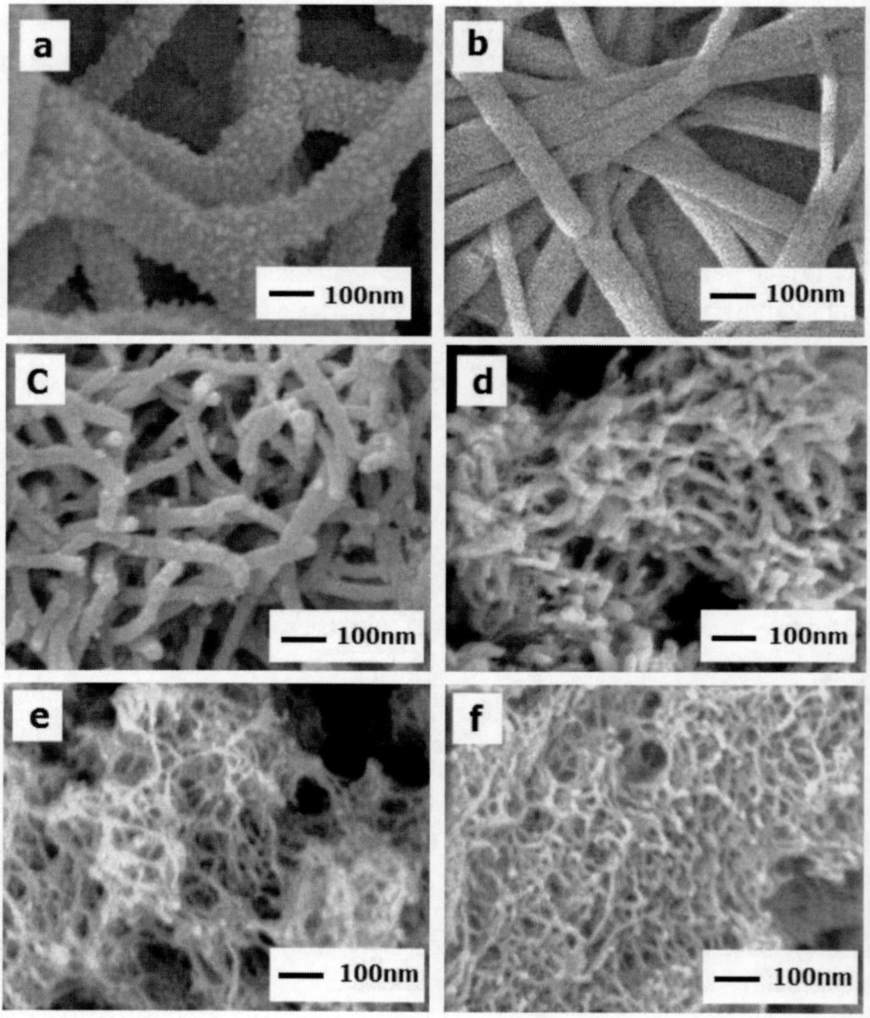

Figure 4. SEM images of the PANI nanofibers prepared by TFM in the presence of different oxidants via chemically oxidative polymerization: (a)[APS]/[An] = 1, the average outer diameter is ~130 nm; (b) [H_2O_2]/[An] = 10, the average outer diameter is ~ 85 nm; (c) [$Ce(SO_4)_2$]/[An] = 2, the average outer diameter is ~ 45 nm; (d) [$FeCl_3$]/[An] = 3, the average outer diameter is ~18 nm; (e) [$Fe_2(SO_4)_3$]/[An] = 4, the average outer diameter is ~13 nm; (f) [$CuCl_2$]/[An] = 6, the average outer diameter is ~12 nm. Synthetic conditions: [An] = 0.1M, reaction temperature: 0~5 °C; reaction time: 15 h.

Table 1. Effect of the redox potential of oxidants on the diameter and conductivity of the PANI nanofibers prepared by TFM via chemically oxidative polymerization

Oxidant	Oxidant/An	Found diameter (nm)	Cal. Diameter [a] (nm)	σ (S/cm^{-1})	Doping degree $[N^+]/[N]$	E_{ox}[18] $H_2/2H^+$
$(NH_4)_2S_2O_8$	1	130 ± 30	128	1.8×10^{-2}	0.24	2.05
$Ce(SO_4)_2$	2	45 ± 10	51	4.6×10^{-2}	0.27	1.44
$FeCl_3$	3	18 ± 5	17	4.4×10^{-2}	0.28	0.77
$Fe_2(SO_4)_3$	4	13 ± 2	15	2.6×10^{0}	0.38	0.68
$CuCl_2$	6	12 ± 3	12	2.2×10^{-5}	0.18	0.56

Polymerization parameters: reaction temperature: 0~5°C; reaction time: 15 h.
[a] : the diameter of resultant nanofibers calculated based on the formula (*).

Based on the standard redox potential of the selected oxidants, the optimum molar ratio of oxidant to aniline (represented by [oxidant]/[An]) was established to be 1:1, 2:1, 3:1, 4:1, 6:1 for [APS]/[An], $[Ce(SO_4)_2]/[An]$, $[FeCl_3]/[An]$, $[Fe_2(SO_4)_3]/[An]$ and $[CuCl_2]/[An]$, respectively. Figure 4 shows typical SEM images of the PANI prepared by template-free method (TFM) in the presence of different oxidants. As one can see, all resulting PANI are fibrous morphology in shape and the fibril morphology is almost independent upon the redox potential of the oxidants. However, the diameter of the fibers significantly decreases with decrease of the redox potential of the oxidants, for instance, the average outer diameter decreases from ~130 nm at E_{ox} = 2.05 V (i.e. APS) to ~12 nm at E_{ox} = 0.56 V (i.e., $CuCl_2$). The diameter of the nanofibers oxidized by different oxidants is listed in Table 1.

Interestingly, there is a well-fitting linear correlation between the standard redox potential of the oxidants and the logarithm of the average outer diameter (d) as shown in Figure 5, which can be expressed by an equation 1 as follows:

$$\log d = 0.69 + 0.71E_{ox} \quad (1)$$

where d is the average outer diameter of the resultant nanofibers and E_{ox} is the standard redox potential of the oxidants respectively. The formula indicates that well–controlling diameter for the PANI nanofibers can be realized by only changing the redox potential of the used oxidants.

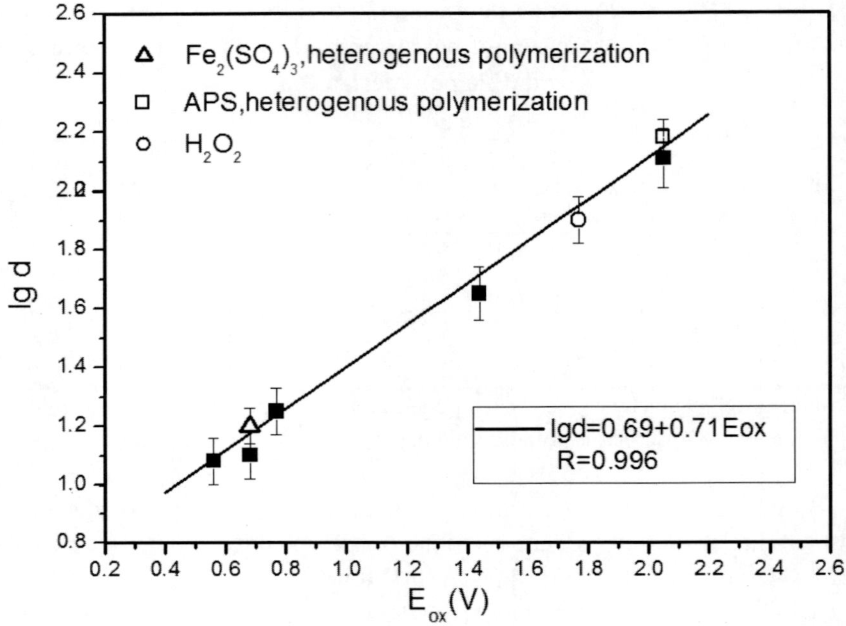

Figure 5. Effect of the redox potential of the oxidants on the diameter of the nanofibers prepared by template-free method (TFM) via chemically oxidative polymerization.

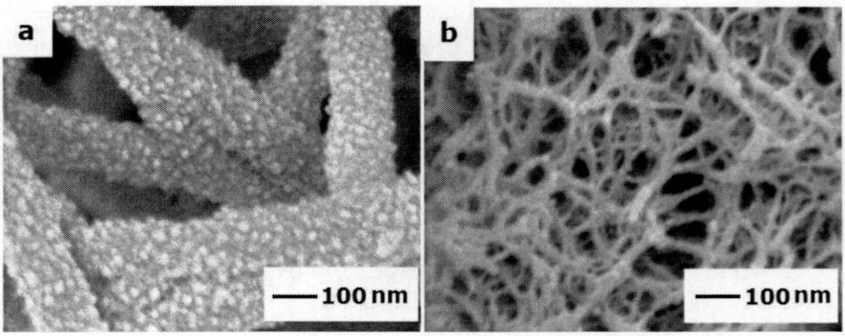

Figure 6. SEM images of the PANI nanofibers prepared by TFM via a heterogeneous polymerization in the presence of different oxidants (a) APS as the oxidant and [APS]/[An] = 1, the average outer diameter is ~150 nm; (b) $Fe_2(SO_4)_3$ as the oxidant and $[Fe_2(SO_4)_3]/[An] = 4$, the average outer diameter is ~15 nm. Synthetic conditions: reaction temperature: 0~5℃; reaction time: 15 h; the volume of water: 5 ml; the volume of the organic phase (chloroform): 10 ml.

In order to confirm the availability of the formula, H_2O_2 (E_{ox} =1.77 V) [44] was chosen as an oxidant to prepare PANI by template-free method

(TFM). It is found that the PANI oxidized by H_2O_2 is also fibrous morphology (see Figure 4b) in shape and its diameter measured by SEM is about 80~90 nm (see Figure 5). This is in good agreement with 87 nm calculated based on the formula, indicating the formula is also suitable for H_2O_2 oxidant. In order to further prove universality of the formula, APS (E_{ox} = 2.05 V) [44] and $Fe_2(SO_4)_3$ (E_{ox} = 0.68 V) [44] were chosen as the oxidants to prepare PANI by TFM via a heterogeneous interfacial polymerization (see supporting information). As shown in Figure 6, the nanofibers oxidized by APS and $Fe_2(SO_4)_3$ are also obtained, and their average diameter of ~150 nm for APS (see Figure 6) and ~15 nm for Fe_2 $(SO_4)_3$ (See Figure 6) measured by SEM also fit the formula, respectively.

In particular, all diameters of the PANI nanofibers prepared by template-free method in the various organic functionalized acids as the dopant [43, 45], for instance, 17~30 nm for $FeCl_3.6H_2O$ and ~150 nm for APS as the oxidant, are well agreement with the above-formula, definitely demonstrating the validity of the formula proposed by us. In particular, a similar relationship is also found in polypyrrole (PPY). By using this formula, thus the diameter of the nanofibers oxidized by a given oxidant can be quantitatively predicated.

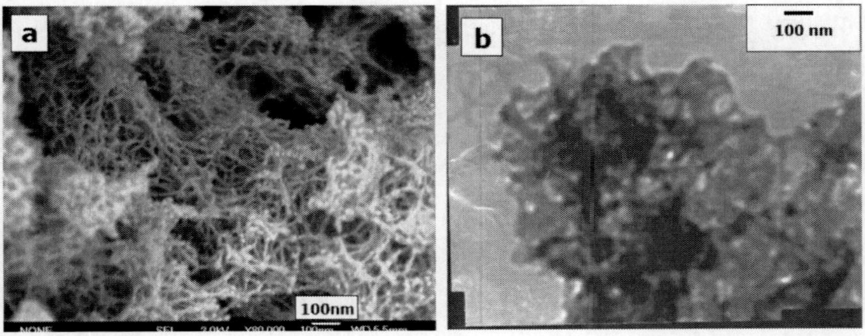

Figure 7. Influence of redox potential of oxidant on morphology and diameter of the PANI nanowires: SEM (a) and TEM (b) of the nanowires oxidized by $Fe_2(SO_4)_3$, respectively. Other reaction condition: [An] = 0.1 M, [APS]/[An] = 1:1; $[Fe_2(SO_4)_3]$/ [An] = 3:1; reaction time :15 h, reaction temperature : 0~5°C.

In principle, aniline monomer can be regarded as amphiphilic molecule due to its hydrophobic group of benzene ring and hydrophilic $-NH_2$ group. As a result, the micelle composed of aniline monomer is expected to be existence in either homogenous or heterogeneous polymerization. The micelles can be served as the soft-templates in the formation of the PANI nanofibers prepared by TFMDF. Once adding the oxidant, the polymerization takes place at the interface of micelle/water due to hydrophilic oxidants [46] and the growth of the nanofibers is controlled by polymerization and elongation process.[47] As above-described, the amount of aniline in this study is remained a constant, while the [Oxidant]/[An] ratios are changed according to the redox potential of the oxidants. Thus the size of the micelles as the soft-templates is expected as a constant, in other word, it is independent upon the redox potential of the oxidants. However, the redox potential of the oxidants will affect the polymerization or elongation process, which controls growth of the nanofibers. It means that the higher redox potential leads to higher polymerization or elongation rate, resulting in a large diameter of the nanofibers for a given polymerization time. This might be why the diameter increases with increase of the redox potential of the oxidants (see Figure 7). As we know, nanomaterials has special characteristic size effect, which means that the conductivity increase with decrease of the diameter [48]. Based on this equation, author chose $Fe_2 (SO_4)_3$ with low redox potential of 0.68 eV as oxidant to polymerize aniline and successfully prepared PANI nanofibers with 10 nm in average diameter, which is the thinnest PANI nanofibers via template-free method, as shown in Figure 7.

To our best knowledge, it is the thinnest nanofibers synthesized by template-free method. Interestingly, the maximum conductivity is achieved as high as ~10^0 S/cm that is enhanced by 10^2 times compared with a conductivity of 8.2×10^{-2} S/cm of the nanowires (150 nm in diameter) oxidized by APS. The higher conductivity of the PANI nanowires at room temperature might result from the size effect [49] caused by the thinner diameter (~10 nm).

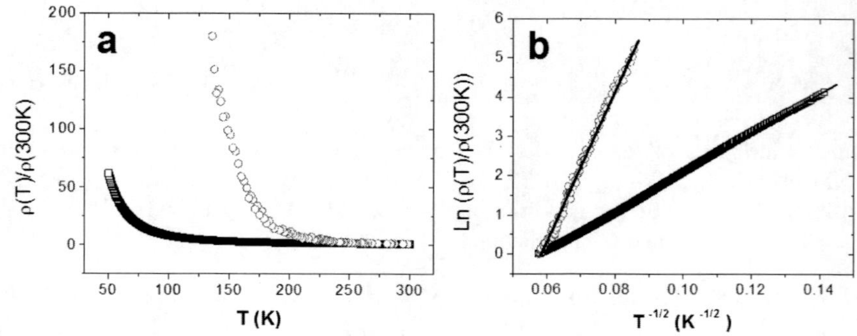

Figure 8. Temperature dependence of resistivity for PANI nanowires oxidized by $Fe_2(SO_4)_3$ (black square, $[Fe_2(SO_4)_3]/[An] = 3:1$) and APS (red circle, [APS]/[An] = 1:1): (a) ρ(T)/ρ(300K) vs. T; (b) ln[ρ(T)/ρ(300K)] vs. $T^{-1/2}$.

Temperature dependence of the conductivity for both PANI oxidized by $Fe_2(SO_4)_3$ and APS was also measured by a four-probe method as shown in Figure 10, where the resistivity was normalized by the resistivity at 300 K represented as ρ(300K). It is obvious that both resistivity increases with temperature decreasing, exhibiting a semi-conductor behavior. However, the nanowires oxidized by $Fe_2(SO_4)_3$ exhibit a much weaker temperature dependence than that of nanowires oxidized by APS. For example, the resistivity ratio of ρ(135K)/ρ(300K) is only 4.1 for the $Fe_2(SO_4)_3$-oxidized nanowires, but 135.2 for the APS-oxidized PANI wires. Further analysis indicates that their temperature dependences of resistivity can be interpreted in terms of quasi-one-dimensional Mott variable range hopping (quasi-1D Mott VRH) model [50], as shown in Figure 8:

$$\rho(T) = \rho_0 * \exp(T_0/T)^{1/2} \quad (2)$$

Where $T_0 = 18 / L_c^3 N(E_F) k_B$ is the characteristic Mott temperature, which can be obtained from the ln ρ(T) ~ $T^{-1/2}$ plot, L_c is the localization length of charge carriers, k_B is the Boltzmann constant and $N(E_F)$ is the density of states at the Fermi level. The value of T_0 is an important parameter, which is related with the sample crystalline and the energy needed for charge carriers' hopping conduction. Usually, the smaller value of T_0 indicates the higher electrical conductivity of the sample [51]. For the present case, the characteristic Mott temperature, T_0, calculated from Figure 8b is 2.5×10^3 K and 3.7×10^4 K for the PANI nanowires oxidized by $Fe_2(SO_4)_3$ and APS, respectively. This result is also consistent with the above results that the $Fe_2(SO_4)_3$-oxidized nanowires show higher room-temperature conductivity and weaker temperature dependence of resistivity.

4. Mono-Dispersed and Oriented CP Nanostructures by Template-Free Method

Mono-dispersed and/or oriented conducting polymer micro/nanostructures have attracted more and more attention due to their highly desire for applications ranging from chemical and biological sensors and diagnosis to energy conversion and storage, light-emitting display devices, catalysis, drug delivery, separation, microelectronics, and optical storage [52].

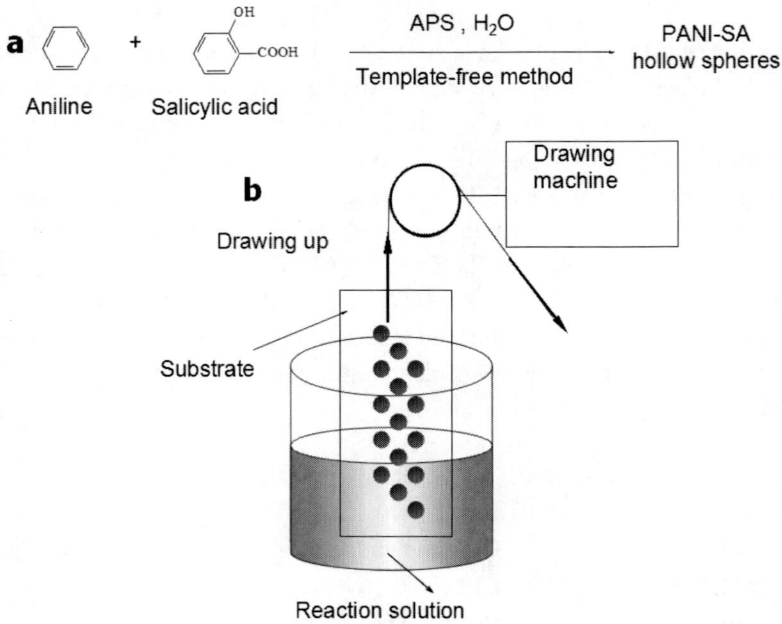

Figure 9. (a) Reaction mechanism sketch and (b) setup illustration for preparation of PANI-SA microspheres via template-free method.

Synthesis or preparation of highly oriented arrays of the micro-/nanostructured conducting polymer is a challenge to chemists and material scientists because they are fundamental conditions to fabricate micro-/nanodevices. As we know, template-synthesis method is one effective and controllable route to fabricate oriented arrays of the conducting polymer micro-/nanostructures. In this approach, however, an external solid material, including anodic aluminum oxide [53], "track-etch" membrane [54], channels of porous alummosiliate [55], or colloids [56] is required as supporting template and subsequently the template is required to be removed to leave the desired polymer micro-/nanostructures that replicate the size and the ordering of the template. However, the removal of the template is rather tedious and this post-processing often destroys product structure. On the other hand, template-free method is simple and cheap resulting in omitting the post-treatment due to lacking of membrane as hard-templates. However, a myriad of nano-scale morphologies can be constructed through self-assembly, but controlling the monodispersed and oriented arrays of nanostructures is quite difficult. Combining template-free method with a deposition process or hard-templates, monodispersed and oriented arrays of PANI could be successfully prepared. Liu and co-workers [57] reported large arrays of uniform and oriented nanowires of conducting polymers on a variety of substrates (Pt, Si, Au, carbon, silica) prepared by the electrochemical synthesis without using a supporting template, which provides a new opportunity for designing devices on the basis of electropolymerization. Therefore, searching a simple and effective method to synthesize or prepare monodispersed and/or oriented arrays of the micro-/nanostructured conducting polymers still remains a scientific challenge. Qiu et al. [58] attempted to obtain oriented nanowires by using self-assembly process associated with a hydrophilic porous Al_2O_3 template. It is found that the nanowires grow vertically from the surface of the porous Al_2O_3 templates to form a 1D highly orderly arrays.

4.1. Mono-Dispersed and Oriented PANI Microspheres

PANI is unique among conducting polymers and has been most extensively studied due to its low cost, easy preparation, controlled unique properties by oxidation and protonation state, excellent environmental stability, as well as its potential application in electronic devices. Recently, three-dimensional (3D) spheres of conducting polymers are also interesting in the fields of material science and catalysis [59]. Therefore, it is an interesting project to prepare or synthesize mono-dispersed and/or oriented arrays of PANI micro-/nanospheres via a self-assembly process. Author successfully used a simple and inexpensive method, a template-free method associated with a deposition process, to prepare the highly oriented arrays and well mono-dispersed PANI spheres with a diameter of 300 ~ 600 nm. Setup used for the deposition of PANI micro/nanospheres doped with SA as the dopant by template-free method associated with a deposition process is shown in Figure 9 (setup).

Typical preparation process for the microspheres of PANI-SA is described as follows: A pre-cleaned glass substrate and treated with a piranha solution is dipped perpendicularly into mixture solution of aniline and SA for a predetermined time. Oxidant (APS) dissolved in aqueous solution is then added into above mixture. The polymerization takes place immediately after adding oxidant, and a thin layer of the doped PANI (e.g. PANI-SA) is formed on the surface of the substrate. The coated PANI substrate is then withdrawn vertically from the suspension at a constant rate Finally the PANI coated substrate is washed with enthanol and copious amount of water several times and dried with N_2 gas at the room temperature.

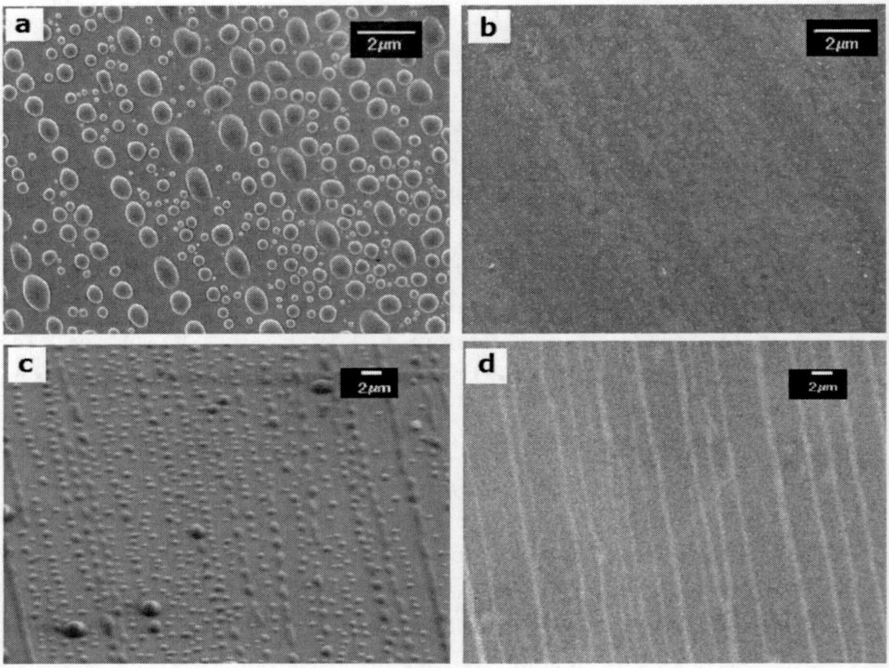

Figure 10. SEM images of (a) PANI-SA spheres deposited on the silicon wafer and (b) silicon wafer; (c) PANI-SA spheres deposited on the PTFE modified glass substrate and (b) PTFE modified glass substrate. Other reaction conditions: lift rate: 1.4 mm/s; polymerization time: 30 min.

Mono-dispersed PANI micro-spheres with 300~400 nm in diameter deposed on the glass substrate were successfully prepared, as shown in Figure 10. Based on the micelles modal [60], micelles composed of aniline and SA might acts a soft-template in the formation of the spheres deposited on the substrates due to its hydrophilic –OH and –NH groups and hydrophobic benzene ring groups. It was found that no any layers were formed on the substrate if the substrate was inserted into the mixture after the color of the reaction solution began green, that was the polymer had been formed, proving it was the micelles that were absorbed on the substrate. As the substrate was immersed into the solution, therefore, it was expected that some micelles were adsorbed onto the hydrophilic solid/liquid interface from the suspension and the diameter did not change once the adsorption occur [44]. During the adsorption process, the hydrophilic group of aniline and SA (-OH, -NH) might easily bond with the H^+ in the surface of the substrates caused by treatment of H_2SO_4-H_2O_2 solution to form mono-dispersed spheres on the substrates.

In case of the oriented substrates (e.g. silicon wafer and glass slid modified with PTFE), those mono-dispersed spheres are favorable to deposit along the channels on the surface of the substrates to form oriented arrays of the spheres due to a low surface energy on the channels. Therefore, as shown in Figure 10a, the highly oriented arrays of the spheres deposited on the (100)-oriented silicon wafer were obtained, which is consistent with the regular silicon channels, demonstrating that self-assembled oriented arrays of the spheres on the substrates can be induced by the oriented surface structure of the substrates. At the same time, it was also confirmed by a fact that uniform mono-dispersed and oriented arrays of the spheres deposited on the glass slid modified with polytetrafluoroethylene (PTFE) was successfully obtained as shown in Figure 10c.

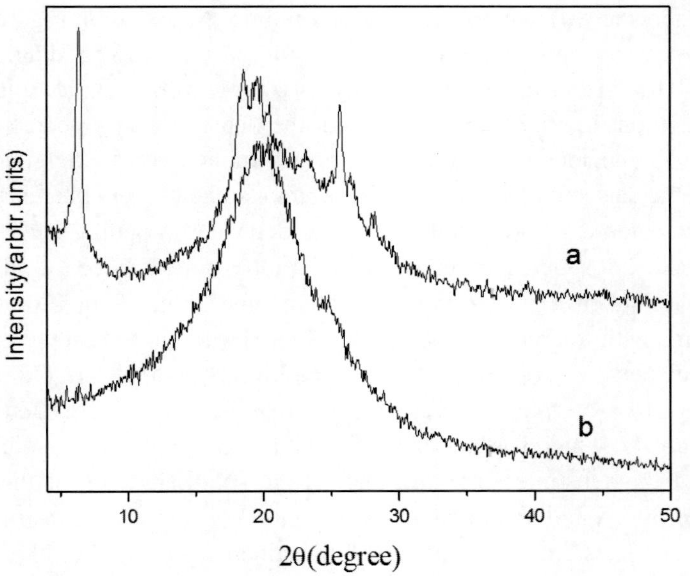

Figure 11. XRD of the PANI-SA spheres: (a) highly oriented arrays deposited on the silicon wafer and (b) the bulk spheres.

Moreover, a significant difference in XRD between the highly oriented spheres monolayer and bulk spheres was also observed as shown in Figure 11. The spheres formed in bulk-solution are amorphous [61], only one broad peak centered at $2\theta = 21^{\circ}$ was observed, which is assigned to planes parallel to the polymeric chain [62]. Except for a broad band at $2\theta = 21^{\circ}$, on the other hand, a sharp peak at $2\theta = 6.46^{\circ}$, which is ascribed as the periodic distance between the dopant and N atom on adjacent main chains [63], is observed from the highly oriented spheres monolayer. Above-mentioned results indicate that the orientation of the spheres deposited on the substrate results in a short-range order along the polymer chain and a highly crystallinity compared with bulk-polymerized spheres.

4.2. Mono-dispersed and Highly Hydrophobic PANI Microspheres

Using the template-free method associated with deposition process, the authors also successfully prepared the hydrophobic surface constructed by PANI microspheres. It is found the diameter and aggregated structures of the PANI spheres are affected by the deposition time. As shown in Figure 12, only few uniform spheres with 300 ~ 400 nm in an average diameter were observed, for instance, when the polymerization time was 15 min. As the polymerization time increased to 30 min, the density of the spheres deposited on the substrate increased greatly to form a monolayer of the spheres fully covering the surface. In the case of the 60 min, some bigger spheres with 2 ~ 5 μm in diameter were adsorbed on the pre-deposited monolayer composed of mono-dispersed spheres to construct a unique co-existence of micro- and nano-scaled spheres. As the polymerization time longer than 120 min, more big spheres with a large diameter of 2 ~ 10 μm were grown on the monolayer of well-dispersed spheres to form a multilayer coated on the substrate. Moreover, mostly microspheres bonded together and aggregated to form disorder arrays, which was similar to that obtained from the bulk-solution [64].

When the substrate was immersed in the solution, some of the micelles composed by SA and aniline might adsorb onto the surface of substrate and act as "soft-templates" in the formation of the spheres deposited on the substrate. According to Biggs′ reports [65], the diameter of the micelles was insensitive to the reaction conditions once the micelles are adsorbed at the solid/liquid interface. However, the micelles remained in the solution could couple with the aniline monomer to grow bigger micelles by polymerization and accretion process [66] under the driving force caused by the hydrogen bond between the –OH group of SA with the amine of PANI. In the initial stage, the micelles with small diameter firstly absorbed onto the surface from the suspension prior to bigger micelles because the heavy mass of the big micelles retarded their movement in the solution and hence their adsorption on the interface. With the polymerization time increasing, big spheres were gradually adsorbed on the surface. It is also found that the water contact angle date is dependent on the polymerization time. As shown in Figure 13, at the polymerization time of 60 min, changed from 15 to 60 min. However, it decreased from 148.0° to 88.7° at the time longer than 120 min. It is clear to show the maximum CA value calculated is about 148.0° at 60 min, exhibiting a highly hydrophobic surface. Combined with morphology shown in Figure 14, co-existence of micro- and nano-scaled spheres, which is rough enough to trap air in the vacancies between individual nanostructures so that the contact area between water and the film can be minimized, resulting in a highly-hydrophobic surface. In general, the CA can be described by Cassie's Equation 3 [67]:

$$\cos\theta' = f_1 \cos\theta - f_2 \quad (3)$$

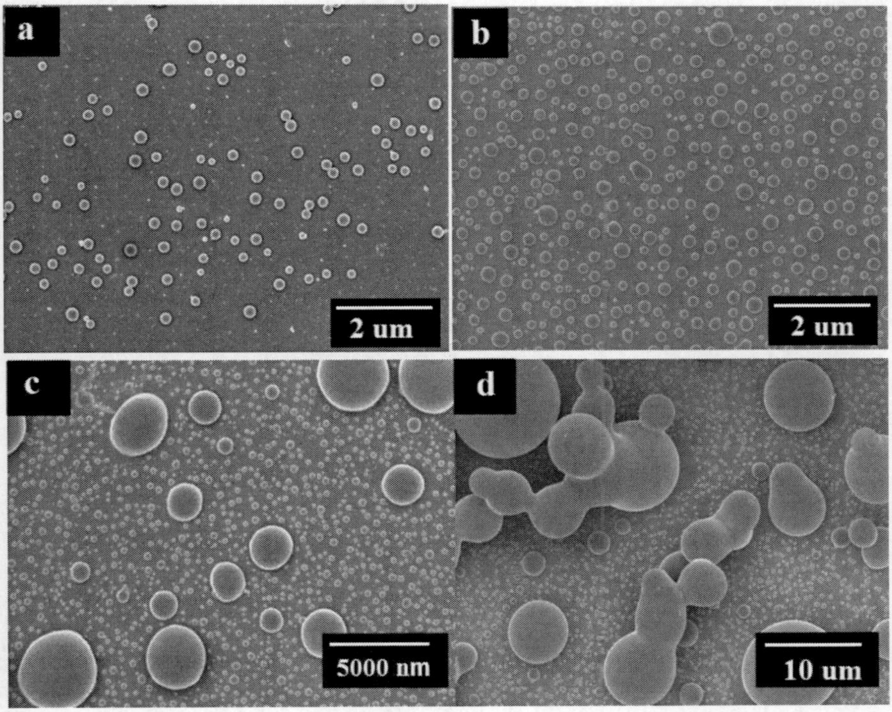

Figure 12. SEM images of PANI nanospheres and microspheres at different deposition time: (a) 15 min; (b) 30 min; (c) 60 min; and (d) 120 min.

Where θ is the intrinsic CA on an ideally smooth surface, while θ′ is the equilibrium contact angle on a rough screen made of the same material as the smooth surface; f_1 and f_2 are the fractions of solid/water interface and air/water interfaces, respectively (i.e., $f_1 + f_2 = 1$). From Equation 3, it is clearly that θ′ increases with f_2, in other words, rough surface implies high fraction of f_2, resulting in hydrophobicity of the surface. Since the sum of f_1 and f_2 is equal to 1, thus the f_2 value for different deposition time can be calculated according to the Equation 3. For example, f_2 was calculated to be 0.91 at θ′= 148.0°, which reveals that highly-hydrophobic surface results from a large fraction of air on the PANI-SA spheres deposited on the substrate.

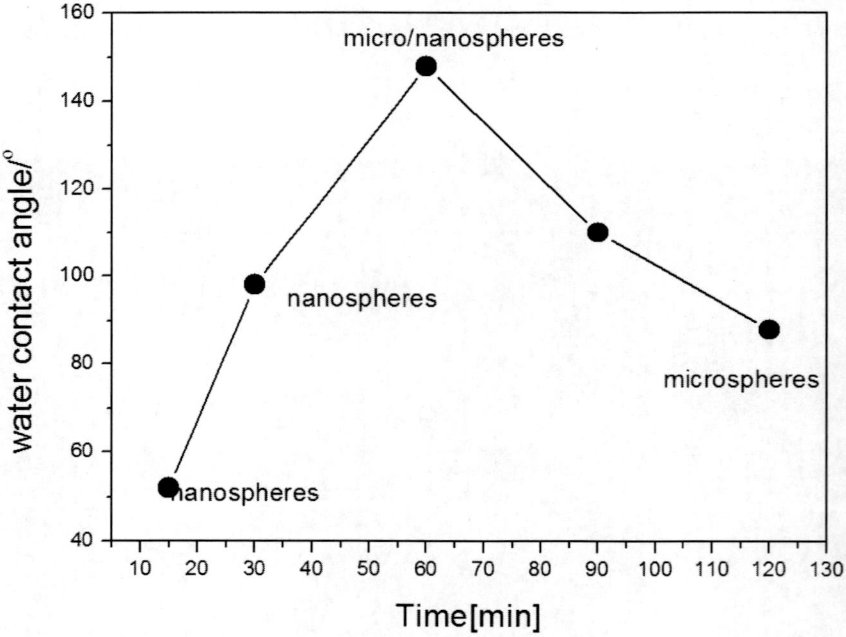

Figure 13. The water contact angle date is dependent on the polymerization time.

Conclusion

"Template-free" method is a facile and efficient approach to prepare conducting polymer (CP) micro-/nanostructures. The formation and growth of CP micro-/nanostructures is a self-assembly process. Molecular interactions such as hydrogen bonds, Van der Waals forces and stacked interactions are usually served as powerful driving forces for self-assembly of micelles, which can act as "soft-template" in the formation of the nanostructures. Compared with template-guided method, it is simple and inexpensive because of omitting template and post-treatment of removing template. By changing the dopant structure and reaction condition, the formation yield, morphology and diameter of the resultant CP micro-/nanostrucutres can be controlled. In particular, it is found that well-controlling diameter of PANI nanofibers can be synthesized by adjusting redox potential of the oxidants. The thinnest PANI nanofiber (~10 nm in diameter) with improved conductivity is successfully synthesized using $Fe(SO_4)_3$ as dopant. Moreover, combination of template-free method with other association approach, such as deposition polymerization, can be used to prepare highly oriented and super-hydrophobic film of PANI micro-/nanospheres.

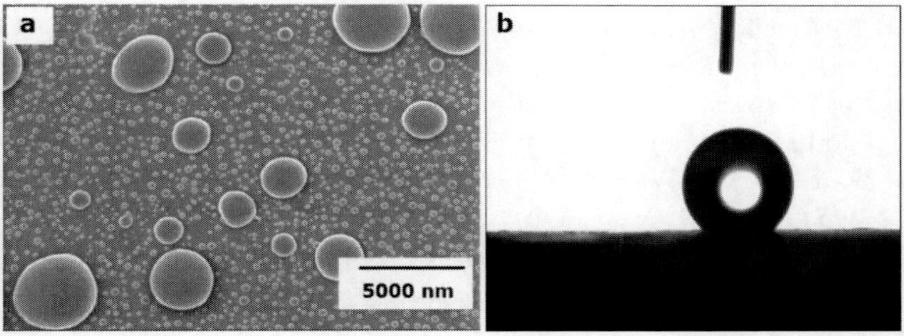

Figure 14. (a) SEM image of co-existence of micro- and nano-scaled PANI spheres; (b) Optical image of a drop of water on the surface coated with micro- and nano-spheres.

ACKNOWLEDGMENTS

This work was supported by the National Natural Science Foundation of China (NSFC51273022 and 51373082), Natural Science Foundation of Shandong Province, China for Distinguished Young Scholars (JQ201103), Taishan Scholars Program of Shandong Province (ts20120528), National Key Basic Research Development Program of China (973 special preliminary study plan, 2012CB722705), and Program for Scientific Research Innovation Team in Colleges and Universities of Shandong Province.

REFERENCES

[1] M. X. Wan, Y. Q. Shen J. Huang, Chinese Patent, No 98109916.5 (1998).

[2] C. R. Martin, *Science*, 266, 1961 (1994).

[3] M. X. Wan, J. L. Li. *Polym. Adv. Tech.*, 14, 320 (2003).

[4] Y. Q. Shen, M. X. wan, *J. Polym. Sci. Part A: Polym. Chem.*, 35, 3689 (1997).

[5] J. Liu and M. X. Wan. *J. Mater. Chem.*, 11, 404 (2001).

[6] Y. S. Yang, M. X. Wan. *J. Mater. Chem.*, 11, 2022 (2001).

[7] Y. S. Yang, J. Liu, M. X. Wan. *Nanotechnology*, 13, 771 (2002).

[8] A. K. Boal, F. Ilhan, J. E. DeRouchey, T. T. Albrecht, T. P. Russell, V. M. Rotello. *Nature*, 404, 746 (2000).

[9] Z. X. Wei, Z. Zhang, M. X. Wan. *Langmuir*, 18, 917 (2002).

[10] M. J. Rosen, In: Surfactant and Interfacial Phenomena (2nd ed.). New York: John Wiley & Sons (1989).

[11] K. Huang, M. Wan, Y. Z. Long, Z. J. Chen, Y. Wei. *Synth. Met.*, 155, 495 (2005).

[12] M. X. Wan, J. C. Li. *J. Polym. Sci. Part A: Polym. Chem.*, 38, 2359 (2000).

[13] R. H. Baughman, J. F. Wolf, H. Echhardt, L. W. Schacklette. *Synth. Met.*, 259, 121 (1998).

[14] J. H. Funrhop, W. Helfrich. *Chem. Rev.*, 93:1565(1993).

[15] Z. M. Zhang, Z. X. Wei, M. X. Wan. *Macromolecules*, 35, 5937 (2002).

[16] Z. X. Wei, Z. Zhang, M. X. Wan. *Langmuir*, 18, 917 (2002).

[17] J. Israelachvili, D. J. Mitchell, B.W. Ninham. *J. Chem. Soc., Faraday Trans.*, 2, 72, 1525 (1976).

[18] M. Harada, M. Adachi. *Adv. Mater.*, 12, 839 (2000).

[19] A. G. MacDiarmid, *Angew. Chem. Int. Ed.*, 40, 2581 (2001).

[20] L. H. C. Mattoso, A. G. MacDiarmid, A. J. Epstein. *Synth. Met.*, 68, 1 (1994).

[21] M. Harada, M. Adachi. *Adv. Mater.*, 12, 839 (2000).

[22] Y. Wei, K. F. Hsueh, G. W. Jang. *Polymer* 35, 3572 (1994).

[23] J. C. Chiang, A. G. MacDiarmid, *Synth. Met.*, 13, 193 (1986).

[24] J. Laska, J. Widlarz. *Polymer* 46, 1485 (2005).

[25] W. J. Zhang, J. Feng, A. G. MacDiarid, A. J. Epstein. *Synth. Met. 84*, 119 (1997).

[26] E. T. Kang, K. G. Neoh, T. C. Tan, S. H. Khor, K. L. Tan. *Macromolecules* 23, 2918 (1990).

[27] W. R. Salaneck, B. Liedberg, O. Inganas, R. Erlandsson, I. Lundstrom, A. G. MacDiarmid, M. Halpern, N. L. D. Somasiri. *Mol. Cryst. Liq. Cryst.* 121, 191 (1985).

[28] Z. X. Wei, M. X. Wan. *Adv. Mater.*, 14, 1314 (2002).

[29] L. J. Zhang, M. X. Wan. *Adv. Mater.*, 13, 815 (2003).

[30] L. J. Zhang, M. X. Wan, Y. Wei. *Synth. Met.*, 151, 1 (2005).

[31] H. Qiu, M. Wan, B. Matthews, L. Dai, *Macromolecules*, 34, 675. (2001).

[32] Z. M. Zhang, Z. X. Wei, L. J. Zhang, M. X. Wan. *Acta Mater.*, 53, 1373 (2005).

[33] L. X. Zhang, L. J. Zhang, M. X. Wan. Y. Wei. *Synth. Met.*, 156, 454 (2006).

[34] L. J. Zhang, M. X. Wan. *Nanotechnology*, 13, 750 (2002).

[35] A.G. MacDiarmid, J.C. Chiang, M. Halpen, W.S. Huang, S.L. Mu, N.L. Somasli, W. Wu, S.I. Yaniger, *Mol. Cryst. Lip. Cryst.*, 121, 173 (1985).

[36] Z. Zhang, J. Sui, L. Zhang, M. Wan, Y. Wei, L. Yu, *Adv. Mater.*, 17, 2854 (2005).

[37] I. Kogan, L. Fokeeva, I. Shunina, Y. Estrin, L. Kasumova, M. Kaplunov, G. Davidova, E. Knerelman, *Synth. Met.*, 100, 303 (1999).

[38] Z. Sun, Y. Geng, J. Li, X. Wang, X. Jing, F. Wang, *J. Appl. Polm. Sci.*, 72, 1077. (1999).

[39] P. S. Rao, D. N. Sathyanarayana, S. Palaniappan. *Macromolecules*, 35, 4988. (2002).

[40] A. Yasuda, T. Shimidzu, *Polym. J.*, 4, 329 (1993).

[41] Y. Wang, Z. Liu, B. Han, Z. Sun, Y. Huang, G. Yang, *Langmuir*, 21, 833 (2005).

[42] E. Erdem, M. Karakisla, M. Sacak, *Europ. Polym. J.*, 40, 785 (2004).

[43] L. J. Zhang, M. X. Wan, Y. Wei. *Macromol. Rapid Commun.*, 27, 366 (2006).

[44] *Inorganic and Analytical Chemistry*, W. L. Guo Ed.; Harbin Institute of Technology Press: Harbin (2004).

[45] L. J. Zhang, Y. Z. Long, Z. J. Chen, M. X. Wan. *Adv. Funct. Mater.*, 14, 693 (2004).

[46] J. H. Fuhrhop, W. Helfrich, *Chem. Rev.*, 93, 1565 (1993).

[47] M. Harada, M. Adachi, *Adv. Mater.*, 12, 839 (2000).

[48] Z. Cai, J. Lei, W. Liang, V. Menon. *Chem. Mater.*, 27:366 (1991).

[49] M. Granstrom, O. Inganas, *Polymer*, 36, 2867 (1995).

[50] B. I. Shklovskii, A. L. Efros, Electronic Properties of Doped Semiconductors. In *Hopping Transport in Solids*; M. Pollak, B. I.

Shklovskii, Eds.;Spring-Verlag, North-Holland, Berlin, Amsterdam, (1984 and 1990).

[51] Y. Z. Long, Z. J. Chen, N. L. Wang, Z. Zhang, M. X. Wan, *Physica B*, 325, 208 (2003).

[52] A. G. MacDiarmid, *Rev. Mod. Phys.*, 73, 701 (2001).

[53] C. R. Martin, *Science*, 266, 1961 (1994).

[54] C. R. Martin, *Chem. Mater.*, 8, 1739 (1996).

[55] C. G. Wu, T. Bein, *Science* 264, 1757 (1994).

[56] P. Jiang, J.F. Bertone, V.L. Colvin, *Science*, 291, 453 (2001).

[57] J. Liu, Y. Lin, L. Liang, J. A. Voiget, D. L. Huber, Z. R. Tian, E. Coker, B. Mckenzie, M. J. Mcdermott, *Chem. -Eur. J.*, 9, 605 (2003).

[58] H. J. Qiu, J. Zhai, S. Li, L. Jiang, M. X. Wan, *Adv. Funct. Mater.*, 13, 925 (2003).

[59] F. Caruso, R.A. Caruso, H. Mohwaid, *Science* 282, 1111 (1998).

[60] Z. X. Wei, Z. M. Zhang, M. X. Wan. *Langmuir* 18, 917 (2002).

[61] L. J. Zhang, M. X. Wan, Y. Wei, *Synth. Met.*, 151, 1 (2005).

[62] J. P. Pouget, M. E. Jozefowicz, A. J. Epstein, X. Tang, A. G. MacDiarmid, *Macromolecules* 24, 779 (1991).

[63] Y. S. Yang, M. X. Wan, *J. Mater. Chem.*, 12, 897 (2002).

[64] L. J. Zhang, M. X. Wan, *Adv. Funct. Mater.*, 13, 815 (2003).

[65] S. Biggs, L. M. Walker, S.R. Kline, *Nano Lett.*, 2, 1409 (2002).

[66] Z. M. Zhang, Z. X. Wei, M. X. Wan, *Macromolecules* 35, 5937 (2002).

[67] A. B. D. Cassie, S. Baxter, *Trans. Faraday Soc.* 40, 546 (1944).

In: Advances in Conducting Polymers Research ISBN: 978-1-63463-258-4
Editor: Laura Michaelson

Chapter 3

PREPARATION AND APPLICATIONS OF CONDUCTING POLYMER ULTRATHIN FIBERS BY ELECTROSPINNING

Yun-Ze Long[1,2,*], Gui-Feng Yu[1,3], Miao Yu[1,4], Wen-Peng Han[1], Xu Yan[1] and Bin Sun[1]

[1]Collaborative Innovation Center for Low-Dimensional Nanomaterials & Optoelectronic Devices, College of Physics, Qingdao University, Qingdao, P. R. China

[2]State Key Laboratory Cultivation Base of New Fiber Materials & Modern Textile, Qingdao University, Qingdao, P.R. China

[3]College of Science and Information, Qingdao Agricultural University, Qingdao, P. R. China

[4]Department of Mechanical Engineering, Columbia University, New York, US

ABSTRACT

Electrospinning is a simple, versatile and efficient method to produce one-by-one continuous ultrathin fibers. Due to low solubility and intrinsic brittleness of conducting polymers (CPs), it is not easy to fabricate CP fibers by direct electrospinning. In the past decade, different strategies have been developed in order to solve this problem and improve electrical

* E-mail address: yunze.long@163.com or yunze.long@qdu.edu.cn.

conductivity of electrospun CP fibers. This chapter briefly summarizes and reviews three approaches to fabricate CP ultrathin fibers by electrospinning process, including direct electrospinning of CPs into fibers, co-electrospinning of blends of CPs with other spinnable polymers, and template-assisted synthesis using electrospun fibers as templates. In addition, the potential applications of electrospun CP ultrafine fibers in flexible and stretchable electronic devices, field-effect transistors, supercapacitors, neural electrodes and interfaces, etc. have also been discussed.

1. INTRODUCTION

Conducting polymers (CPs), or conjugated polymers such as polyaniline (PANI), polypyrrole (PPY), polythiophene (PT) and its derivatives of poly(3,4-ethylene dioxythiophene) (PEDOT) and poly(3-hexylthiophene) (P3HT) (Figure 1a) [1,2] have been recognized as promising organic semiconductors due to their small size, large specific surface area, high aspect ratio (length/diameter), light weight, low cost, good biocompatibility, facile processability, and adjustable electrical conductivities from 10^{-11} to 10^{5} S cm^{-1} upon doping (Figure 1b), reversible doping/dedoping process and thus unique electrical, magnetic, mechanical and optical properties. These make 1-dimensional (1D) nanostructures have important applications in electronic and optoelectronic devices, chemical and bio-sensors, artificial muscles and biomedical materials, anti-static and electromagnetic shielding materials, energy harvesting and storage devices, etc.[5-9]

CP nanotubes/fibers can be prepared by a variety of approaches such as electrospinning [3], hard template synthesis [8], interfacial polymerization [9], template-free method [10], dilute polymerization [11], reverse emulsion polymerization [12], and lithography technique [13]. Among various approaches established, each of the methods possesses advantages and disadvantages. However, electrospinning is a highly versatile method for the fabrication of ultrafine fibers from a huge of materials either in the form of individual fiber or non-woven fiber mats with nanometer-to-micrometer size diameters, which are 10^{2} to 10^{4} times smaller than those fabricated by the conventional approaches of solution or melt spinning. Although the process of using electrostatic forces to form synthetic fibers has been known for over 100 years, it was not until 1934, the first patent on electrospinning (a process and an apparatus that used electric charges to spin small-diameter fibers) was applied by Formhals [14]. However, this technology was regained more attention until the work of Reneker's group in the 1990s [15]. Since then, the electrospinning process has drawn surging interest in nanotechnology, advanced nanomaterials and devices.

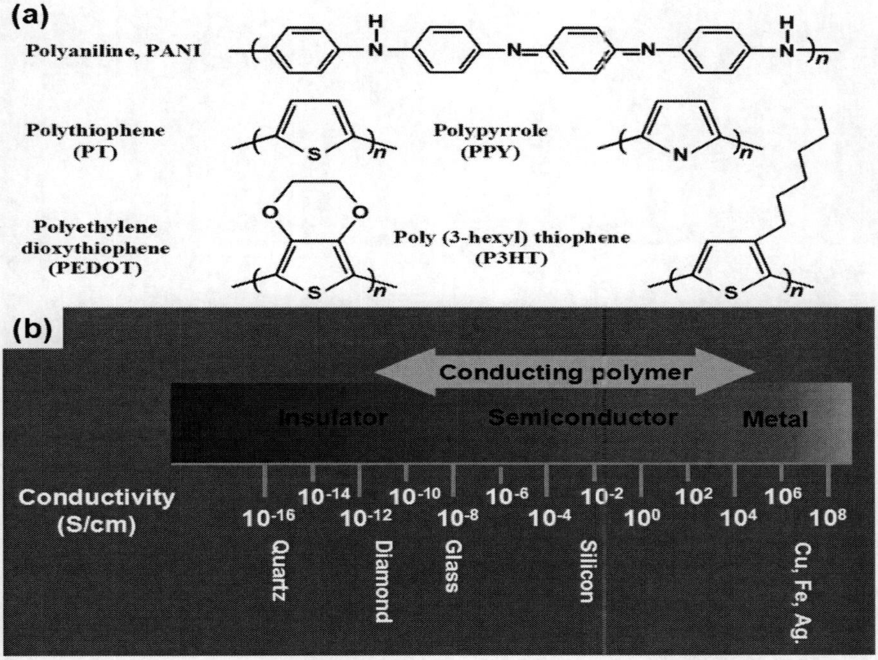

Figure 1. (a) Chemical structures of several CPs [2]. (b) Electrical conductivity range of CPs.

Besides straight fiber morphology, nanoporous structures [16-18], microspheres [19,20], microcups [21,22], ribbons [23], helical fibers [24,25], hollow tubes [17,26-28], multichannel tubes [29,30], 1D colloidal assemblies [31,32], and other structures also can be produced by electrospinning techniques, as shown in Figure 2. For example, core-shell nanofibers and hollow nanotubes could be obtained by coaxial electrospinning and emulsion electrospinning [33, 34]. Moreover, hierarchical multichannel fibrous tubes can be fabricated when multi-capillaries are embedded in a sheath syringe [29]. Figure 2i-j show the as-prepared tubes with four and five channels, respectively. Particularly, the fiber orientation and organization also can be achieved by appropriate selection of electrode configurations or modified electrospinning setups, including parallel pair electrodes collection [35-37], rotating drum or disc collection [38, 39], frame collector [40], step collector

[41], auxiliary electric [42, 43] or magnetic [44, 45] electrospinning, gas-assisted electrospinning [46, 47], centrifugal electrospinning [48], double spinning [49] and electroconductive template collection [50].

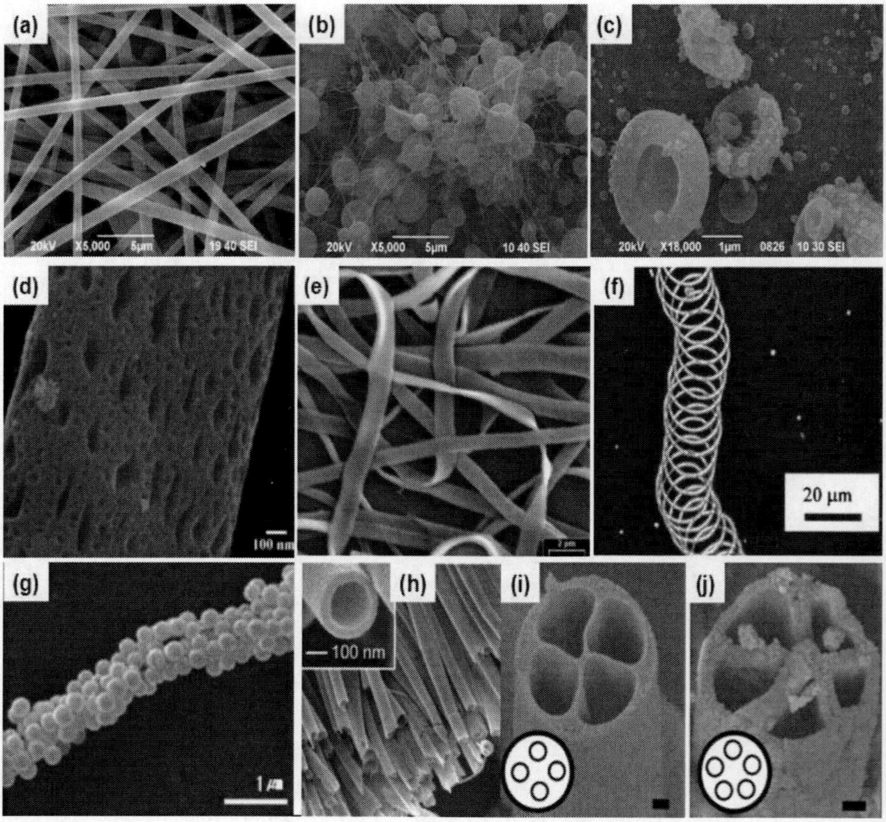

Figure 2. Various morphologies and microstructures of electrospun structures: (a) straight fibers; (b) microspheres; (c) microcups; (d) nanoporous fibers [16]; (e) microribbons [23]; (f) helical fibers [25]; (g) 1D colloidal assembly [31]; (h) hollow fibers [35]; (i) microtube with four channels [29]; (j) microtube with five channels [29].

In order to keep up with this highly active field of nanoscale conducting polymers, therefore, in this chapter, we present a brief summary of recent advances in the synthesis approaches and applications of 1D nanostructures of conducting polymers based on our results and some important contributions of other groups.

2. Strategies for Fabricating of CP Fibers via Electrospinning

Although electrospinning has been recognized as a simple, versatile and efficient technique for production of one-by-one continuous CP ultrafine fibers, the low solubility and intrinsic brittleness of CPs is a major obstacle to produce CP fibers by direct electrospinning. In order to solve this problem and improve electrical conductivity of electrospun CP fibers, different approaches have been developed. In this section, we will summarize various strategies to obtain CP ultrafine fibers by electrospinning process, including direct electrospinning of CPs into fibers, co-electrospinning of blends of CPs with other polymers, and template-assisted synthesis using electrospun fibers as templates.

2.1. Direct Electrospinning CPs into Fibers

In order to keep higher electrical conductivity and other intrinsic physical and/or chemical properties of CPs, in principle, CPs should be electrospun directly just like other polymers. However, due to the limitations on molecular weight and solvents suitable for electrospinning, up to date, only a few soluble CPs such as PANI [15, 51] and PPY [52] could be electrospun directly into fibers. For example, as shown in Figure 3a, Yu and coworkers [53] dissolved high molecule weight HCl/H_2SO_4-doped PANI powder in hot sulfuric acid under magnetic stirring for at least 5 h, and the resultant solution (PANI concentration was 10.6 wt% to 19.1 wt%) could be electrospun into homogeneous PANI sub-microfibers. In this process, a voltage of 12 or 18 kV was applied and the fibers were collected on a coagulation bath, which was made of different concentration sulfuric acid solution ranging from 0 to 30 wt%. The nozzle tip-to-collector distance was 13 cm. After the PANI fibers were dipped in the coagulation bath for 24 h, they were pulled out from the bath and dried at room temperature. The PANI fibers with a diameter of 370 nm exhibited a high conductivity of 52.9 S cm^{-1}. Srinivasan et al. [54] also synthesized PANI fibers using camphorosulfonic acid (CSA) as a surfactant by direct electrospinning method (schematically represented in Figure 3b).

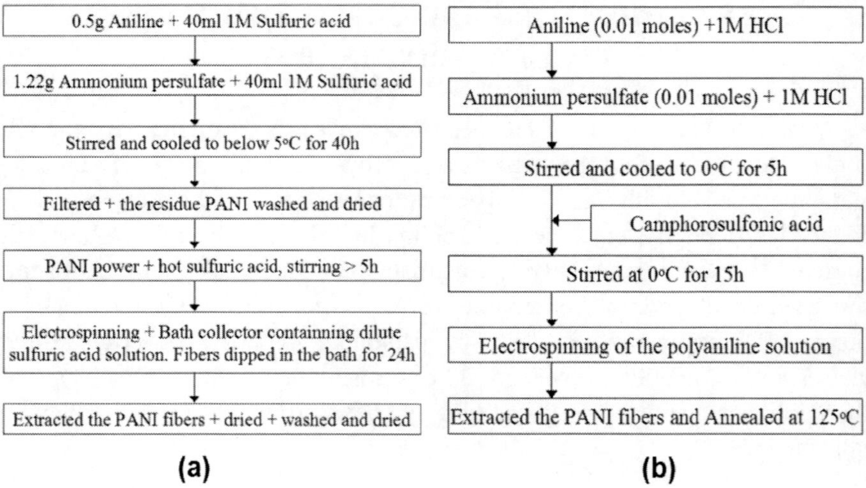

Figure 3. Flow charts for the synthesis of pure PANI ultrafine fibers via direct electrospinning: (a) using a coagulation bath as collector; (b) using camphorosulfonic acid (CSA) as surfactant [54].

2.2. Co-electrospinning of Blends of CPs and Other Polymers

To overcome some drawbacks such as poor mechanical properties, poor processability, hydrophobicity and non-degradability of CPs, most of the researchers electrospun CP through blending it with other spinnable polymers. This strategy can prepare different, various and even multi-functional CP-based composite fibers. However, the electrical conductivity of the as-spun fibers is usually decreased by the additive polymer carrier.

2.2.1. One-Step Co-Electrospinning of CP Blends

In this technique, CPs and spinnable polymers which to be blended should be dissolved in the solvent in sequence in order to obtain the resulting electrospinning solution [55-57]. For example, polymethyl methacrylate (PMMA) was dissolved in N-methyl-2-pyrrolidone (NMP), then PANI were added into the above PMMA/NMP solution. By electrospinning, the PANI/PMMA composite nanofibers could be obtained [55]. Lin et al. used processable PANI as the sensitive material to fabricate nanofiber humidity sensors from the blend of PANI, poly(vinyl butyral) (PVB) and poly(ethylene oxide) (PEO) by electrospinning. The composite nanofibers showed good adhesion to the electrode, good electrical contact, and maintained relatively high specific surface area. Furthermore, the humidity sensors revealed much higher sensitivity than their film counterparts (impedance changing for three orders of magnitude from 20% to 90%RH) [57]. However, the conductivity of the as-spun PANI fibers was relatively lower than that of PANI made by some methods. For instance, MacDiarmid et al. [3] reported that the conductivity was 0.2 S cm^{-1} when electrospun 50 wt% CSA-doped PANI blended with PEO, dissolved in chloroform Which is lower compared with the conductivity of 52.9 S cm^{-1} [53] and 21.8 S cm^{-1} [58] of PANI made by direct electrospinning and solution deposition polymerization, respectively. One of the advantages of this technique is that the property or application of the as-spun composite fibers could be tunable or improved by adding spinnable polymers.

2.2.2. Spinning CP Monomer Blends and Followed in Situ Polymerization

A novel method [59] was introduced to fabricate aligned nano-branched coaxial PANI/PVDF ultrafine fibers via electrospinning as a novel strain sensitive flexible microelectronic material. The conductivity of fibers was about 0.6 S cm^{-1}, which combined the advantages of conducting polymer and elastic material. The aligned PVDF fiber was fabricated by an electrospinning setup with a roller as collector. An eight column roller with the speed of 600 rpm was applied as negative electrode and used to collect aligned PVDF nanofibers. Then aligned PVDF nanofibers was immersed in blended solution and kept steady in the room temperature for 5 h to generate PANI on the surface of the fibers by chemical oxidative polymerization. Finally, the samples were taken out and washed by water for three times and dried in drying case for 30 min at 70 °C. PANI were covered on the surface of the PVDF fibers because of interfacial compatibility of PANI to PVDF. The microstructure of the PVDF fibers and surface features of coaxial fibers were characterized in Figure 4 by scanning electronic microscopy. This special structure increased the content of PANI in this composite fibers to improve the conductivity of the material. In addition, the structure also made this material to have a big specific surface area, so that the material may shows excellent performances in some contact-dependent experiments like gas sensitivity test and so on.

Zhang et al. [60] also used the similar method to fabricate the PEDOT shell/PLGA core fibers with the oxidative catalyst ($FeCl_3$). When the composite fibers were immersed in chloroform to remove the PLGA and any residual EDOT, the hollow PEDOT tubes were obtained. The conductivity of PEDOT fibers ranging from 0.02 to 0.28 S cm^{-1} could be tuned by changing the concentration of PEDOT and $FeCl_3$. The method could be used in supporting and directing the extension of dorsal root ganglia neurons and as a potent drug delivery system.

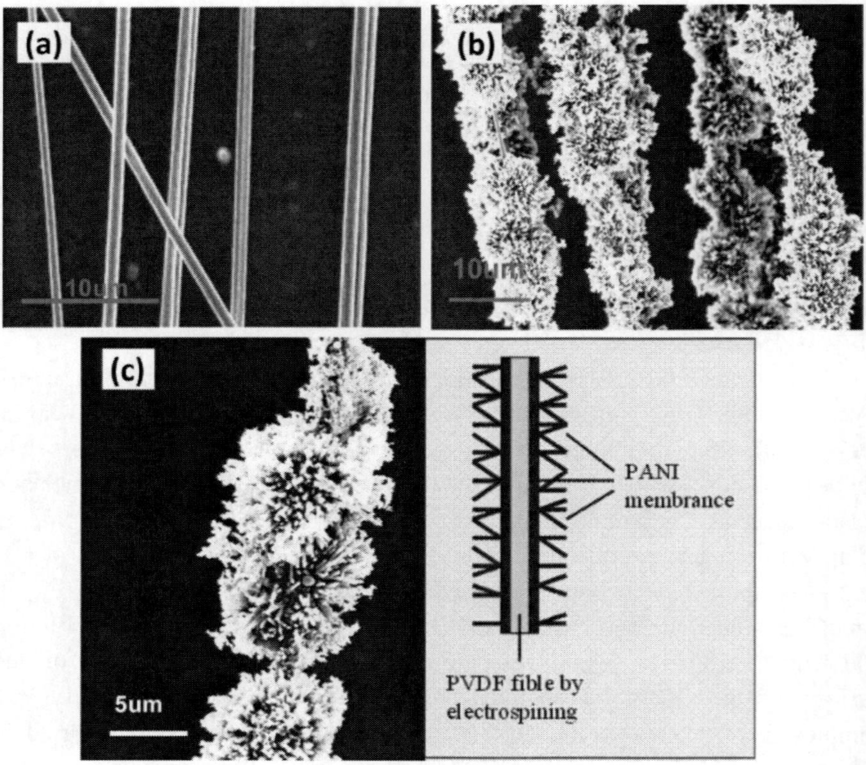

Figure 4. SEM images and the schematic diagram of the materials. (a) SEM image of electrospun PVDFfibers; (b) SEM image of coaxial PANI/PVDF fibers after *in situ* chemical polymerization; (c) SEM images of a single composite fiber and the schematic diagram for its structure.

2.2.3. Coaxial Electrospinning of CPs and Other Polymers

Two-fluid coaxial electrospinning or electrospinning with a two-capillary spinneret is a novel technology to fabricate continuous core-shell CP composite fibers and even pure CP nanotubes. As shown in Figure 5 [34], taking advantage of a coaxial capillary geometry, with one capillary inside another, pumping an electrospinnable solution through the outer capillary, and pumping almost any fluid though the inner capillary, and ultrafine fibers can be produced with a core-shell morphology. Particularly, the core or shell could be removed by a further treatment. For example, poly[2-methoxy-5-(2-ethylhexyloxy)-1,4-phenylenevinylene] / poly(vinyl pyrrolidone) (MEH-PPV /PVP) and MEH-PPV(P3HT)/PVP core-shell fibers were prepared for the first time by electrospinning with a two-capillary spinneret, and conducting MEH-PPV(P3HT) fibers containing different amounts of P3HT could be further obtained by removing the PVP shell through ethanol extraction [61]. Compared with single-capillary electrospinning, the molecular orientation of CPs is developed with the help of spinnable shell-polymer flow, which leads to a more stabilized elongation process and a slower solidification process. Thus, the CPs core flow will stretch along the shell-polymer axis. However, the adventitious doping such as air oxidation and moisture which can be regarded as spontaneous carrier is blocked due to the existence of shell-polymer which may result in a lower conductivity.

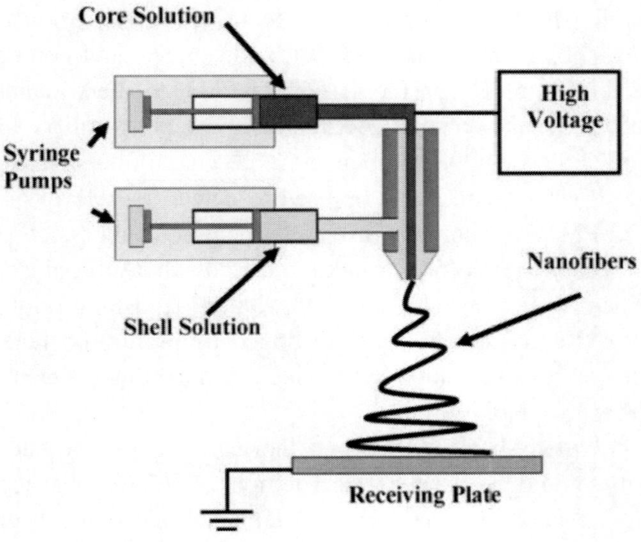

Figure 5. Schematic diagram of a set-up for coaxial electrospinning. [34].

2.3. Electrospun Fiber-Template-Assisted Synthesis

To overcome the disadvantages of direct electrospinning CPs and co-electrospinning CP blends mentioned above, template-guided synthesis using electrospun nanofibers as template is an effective strategy to fabricate coaxial

CP fibers or CP nanotubes. The as-prepared CPs nanostructures not only possess the intrinsic higher electrical conductivity of CPs, but also enhance physical/chemical properties of composite nanofibers to some extent [62,63]. In this method, electrospun polymer or inorganic nanofibers provide a robust and stable template for the growth of 1D nanostructures of various CPs. Particularly, the nanofiber template could be removed during or after the polymerization process of CPs. Up to date, two approaches are usually used to grow CPs on electrospun fiber templates: one is solution deposition polymerization and the other is vapor deposition polymerization.

2.3.1. Solution Deposition Polymerization

In this approach, nanofiber templates are immersed in the solution containing CP monomers, oxidant, and doping acid. The CP monomers are simultaneously deposited on the template surface to form core-shell composite nanofibers. For example, Tang et al. [58] used the electrospun PMMA fibers as core material which were immersed into the mixed solution containing aniline monomer, oxidant (ammonium peroxodisulfate), and doping acid (5-sulfosalicylic acid) at 5 °C. On the surface of PMMA fibers, nanostructured PANI was grown and the coaxial PANI/PMMA composite fibers with a high conductivity of 2.123 S cm^{-1} were obtained. The coaxial fibers also exhibited a higher gas sensitivity, which could be used to detect trace level of ammonia (80 ppb). PANI/PMMA nanofibers also could be used as a gas sensor towards triethylamine (TEA) vapors [64]. In addition, Coaxial PANI/polyacrylonitrile (PAN) nanofibers were also fabricated by solution deposition polymerization [65]. When PANI/PAN nanofibers were doped by perfluorooctane sulfonic, the resultant nanofibers were superhydrophobic and exhibited a chemical dual-responsive surface wettability [66].

Besides polymer fibers, electrospun inorganic nanofibers such as TiO_2 and Mn_2O_3 also could be used as hard templates [67,68]. Particularly, the inorganic Mn_2O_3 nanofibers may serve as template and oxidant at the same time. For example, Li et al. [68] presented a new method for the synthesis of PANI nanotubes with an average diameter of 80 nm and inner diameter of 38 nm. In this method, Mn_2O_3 nanofibers prepared by electrospinning were used as the reactive/oxidant template to initiate the polymerization of aniline in acid solution. After reaction, PANI shells were formed on the Mn_2O_3 nanofiber surface, and the Mn_2O_3 nanofibers were spontaneously removed. As a result, PANI nanotubes were obtained, as shown in Figure 6. In addition, the PANI nanotube sensing device could detect as low as 25 ppb NH_3 in air at room temperature with good reversibility and the response time was about 50 s at 50 ppb ammonia, which indicated that the as-obtained PANI nanotubes is a promising candidate for gas sensing.

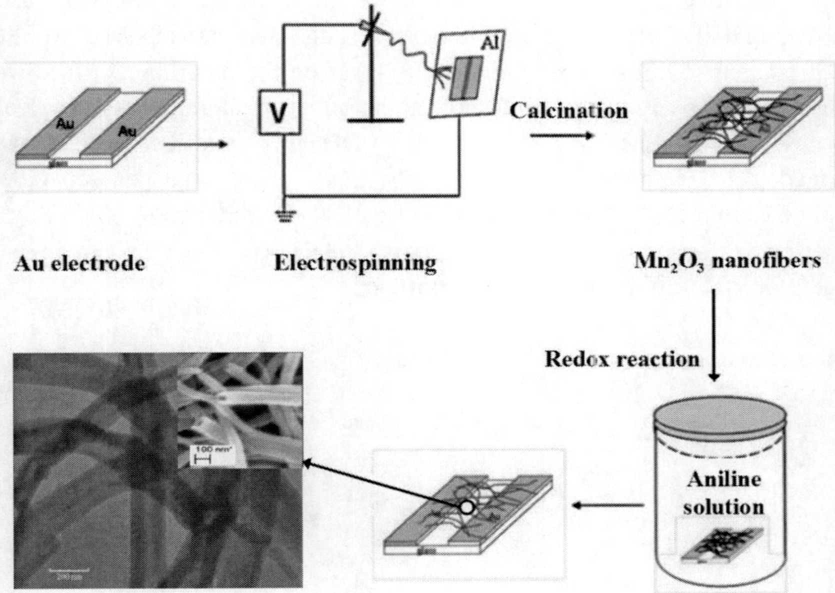

Figure 6. Schematic illustration for the fabrication of PANI nanotubes, scanning electron microscope (SEM) and transmission electron microscope (TEM) images of the PANI nanostructures. [68]

2.3.2. Vapor Deposition Polymerization

In this approach, template nanofibers containing chemical oxidants such as $FeCl_3$ or organic oxidant benzoyl peroxide [69], which are capable of polymerizing CP monomers, are electrospun firstly, and then put into a vessel filled with CP monomer vapor. The monomer vapor is oxidized to CP nanostructures on the surface of the fiber templates. Various CP/polymer core-shell composite fibers have been prepared by this route, including PPY/PEO [70], PPY/poly(vinylidene fluoride-co-hexafluoro propylene) (PVDF/HFP) [71], PPY/polyamide 6 [72], PPY/PS [73], PEDOT/PS [74], etc. For example, Laforgue and Robitaille [75] have fabricated PEDOT fibers by electrospinning and an adapted vapor-phase polymerization procedure. As shown in Figure 7, the PVP nanofibers containing iron(III) p-toluenesulfonate (FeTos) were prepared by electrospinning firstly, and then the nanofibers were put into a glass reactor full of the EDOT monomer vapors by a small vial containing the liquid EDOT at the bottom of the reactor. As a result, the EDOT monomers were polymerized on the surface of the oxidant nanofibers. The PEDOT nanofibers exhibited a high conductivity of 60±10 S cm^{-1} and could be used as sensors and energy storage systems. In addition, Wang et al. [71] used electrospun PVDF-HFP nanofiber membrane as the substrate materials to filter FeCl3 ethanol solution. A semi-dry-$FeCl_3$ layer on the nanofiber surface was formed. By exposure of the $FeCl_3$-coated nanofiber membranes to the pyrrole vapor, a thin coating of PPY on the PVDF-HFP nanofiber surface was obtained. The resultant nanofiber mats have been used as separation membranes to recover Au from aqueous [Au(III)Cl4]- solutions.

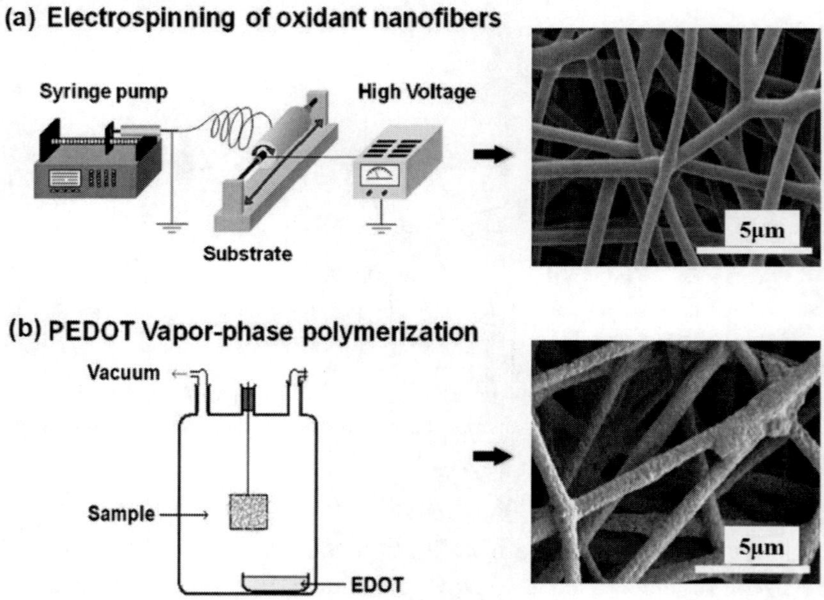

Figure 7. Schematic of the two-step nanofiber production to prepare PEDOT nanofibers. (a) electrospinning equipment and SEM image of PVP nanofibers after electrospinning (b) vapor-phase polymerization system and SEM image of PEDOT/PVP nanofibers after EDOT polymerization. [75].

In addition, as mentioned in the last section, electrospun inorganic nanofibers may serve as template and oxidant at the same time. For example, bamboo-like PPY nanotubes has been successfully fabricated by such a reactive-template vapor phase polymerization method. Firstly, the V_2O_5 nanofibers prepared by electrospinning were chosen as templates to deposit PPY, which served as the oxidants as well. Then the nanofibers were exposed to pyrrole monomers, and through subsequent processing, the PPY nanotubes were obtained. The resulting bamboo-like PPY nanotubes could be used as Cr(VI) ion adsorbents in aqueous solution and exhibited much higher adsorption performance than traditional PPY nanoparticles [76].

2.4. Aligned CP Nanofibers

In recent years, highly aligned nanofibers have become one of the most prominent nanostructures for their applications in electrical, chemical and biomedical fields due to the orientation and improved mechanical properties, etc. For example, the photoelectric devices and gas sensors fabricated by aligned CP nanofibers instead of non-woven fibers usually exhibit a more excellent performance. Especially in the tissue engineering fields such as neurite and muscle outgrowth, the highly aligned fibers can provide a direction guidance in the process of the release and growth of proteins. Recent studies have reported various modified electrospinning techniques in order to prepare highly aligned electrospun fibers instead of non-woven fibers by modifying the exterior conditions of electrospinning including collecting device, spinneret device or outer electric field distribution etc. However, towards aligned CP nanofibers, there are only a few reports about the fabrication including rotating drum/disk collector [77], an alternative electrostatic field generated between two collectors [78], gapping method or two parallel collectors [79], or spinneret writing technique [80]. For example, Ku et al. [77] took advantage of an electrospinning process using a rotating drum collector to fabricate PANI/PCL nanofibers which were suitable scaffold material for tissue engineering. However, the orientation and diameters of the prepared nanofibers were influenced by the rotation speed to a great extent.

To improve the controllability of orientation and positioning of electrospun CP nanofibers, Attout et al. [78] used a novel method based on electrostatic steering to align PANI/PEO and PANI/PMMA blend nanowires on a substrate between two collectors during the electrospinning. While one collector was grounded, the other was maintained at high voltage (6 kV). By means of high voltage reed relays, this electric state was alternated between the two collectors. As a result, the electrostatically charged polymer leaving the capillary bounced from one collector to the other leaving aligned fibers on the substrate.

Except for this method, Yin and his coworkers took advantage of a U-shaped metal device to obtain highly aligned conducting nanofibers via electrospinning [79]. By manually moving the device back and forth for a few times between the spinneret and the electric ground during electrospinning, the highly aligned nanofibers were collected between the metal legs. By this way, the uniaxially aligned electrospun conjugated polymer nanofibers of P3HT/PEO and MEH-PPV/PEO with the highly polarized emission due to the alignment of macromolecular chains were prepared.

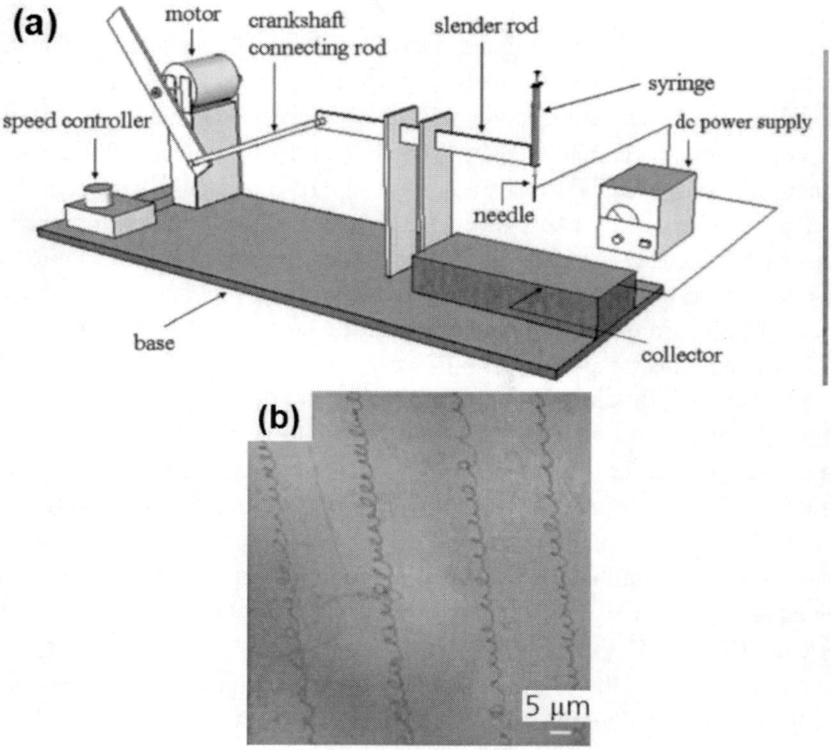

Figure 8. Schematic diagram of reciprocating-type electrospinning apparatus.

In addition, our group has fabricated the stretchable strain sensors based on aligned microfibrous arrays of PEDOT:PSS/PVP with curled architectures by a novel reciprocating-type electrospinning setup with a spinneret in straightly simple harmonic motion (Figure 8). The strain sensor exhibited a linear elastic response to strain, and had a high stretchability to accommodate strains of 4%, and a conductivity of 1.6×10^{-5} S cm^{-1}. In addition, the PEDOT:PSS/PVP microfibrous array strain sensor had a high gauge factor of about 360, as well as a high repeatability and durability [80]. Our group [59] has also invented an approach to fabricate the aligned nanofibers. In this method, we used a rotating disk to collect aligned non-conductive polymer (such as PMMA) fibers firstly, and then the PMMA fibers were immersed into reaction tank which containing CP monomers (such as aniline monomers). After a period of time, the PANI was polymerized on the surface of PMMA fibers and the aligned core-shell PANI/PMMA fibers with a high conductivity of 10^{-1}-10^{2} S cm^{-1} were obtained.

3. Applications of Electrospun CP Ultrafine Fibers

Up to date, ultrafine CP fibers have drawn much attention as promising materials in many fields. Due to their attractive properties associated with conventional polymers, such as adjustable conductivity from 10^{-11} to 10^{5} S cm^{-1}, high specific surface area and fibrous morphology, ease of synthesis and processing, low cost and good environmental stability. In this section, we will briefly introduce some interesting advances in applications.

3.1. Flexible and Stretchable Electronic Devices

An emerging development topic in electronics focuses on stretchable devices for sensing applications including biomedical, food control, demotic and robotics with very light weight, robustness and low cost. It is reported that electrospinning has potential in this field, especially providing flexible and stretchable pressure sensors, strain sensors [80,81-87] except for flexible gas sensors [88]. Here the conductive polymers are dominated because they can exploit deformations induced by small forces, through pressure, mechanical vibration, elongation/compression, bending or twisting [89], and fibrous arrays are able to demonstrate better performance than that non-woven mats or isolated strands in this area. For instance, Our group [80] have fabricated stretchable strain sensors based on aligned microfibrous arrays of poly(3,4-ethylenedioxythiophene): poly(styrene sulfonate)- poly(vinyl pyrrolidone) (PEDOT:PSS-PVP) with curled architectures by a novel reciprocating-type electrospinning setup with a spinneret in straightforward simple harmonic motion. Besides a high room-temperature conductivity of the composite fibers (1.6×10^{-5} S cm^{-1}), and a high repeatability and durability, the sensors could be stretched reversibly with a linear elastic response to strain up to 4% owing to the curled architectures of the as-spun fibrous polymer arrays, which was three times higher than that from electrospun PEDOT:PSS-PVP nonwoven mats. As shown in Figure 9, the electrical properties of the device under bending–releasing remain unchanged which was important for real applications. These results may be helpful for the fabrication of stretchable devices which have potential applications in some fields such as soft robotics, elastic semi-conductors, and elastic solar cells.

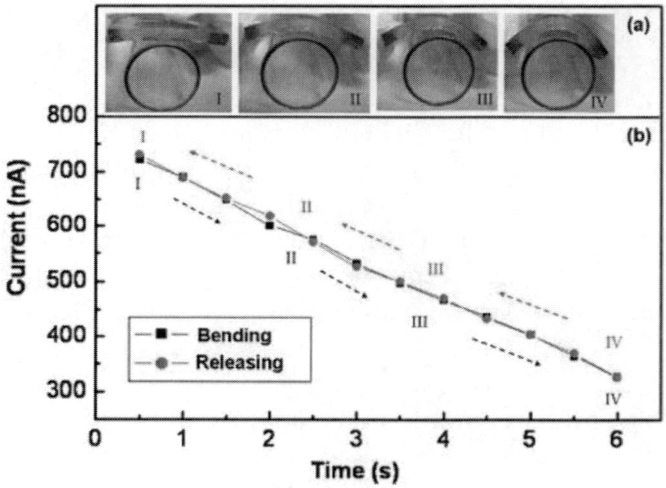

Figure 9. (a) (I–IV) Optical images during the bending process to test the flexibility of the stretchable strain sensor. (b) Current response of this device during the continuous process of bending and releasing.

Through electrospinning and in situ polymerization, our group [59] also fabricated a kind of aligned nano-branched coaxial polyaniline/polyvinylidene fluoride (PANI/PVDF) fibers as a novel strain sensitive flexible micro-electronic material, as shown in Figure 10. Figure 11a gives a strain sensitivity graph character by the relationship of relative resistance change rate and curvature. Figure 11b shows a stable and sensitive response performance of this material in the repeated curving and straightening experiment. The results indicated that the PANI/PVDF microfibers could be used as strain sensor with high flexibility, high sensitivity, and stable repeatability. We believe it would have enormous applications prospects in electronic engineering especially in the field of microelectronic technology for precision measurements, highly sensitive robot sensors such as biologically inspired skin, human-benign devices, personal health monitoring, disaster warning, touch panel and so on.

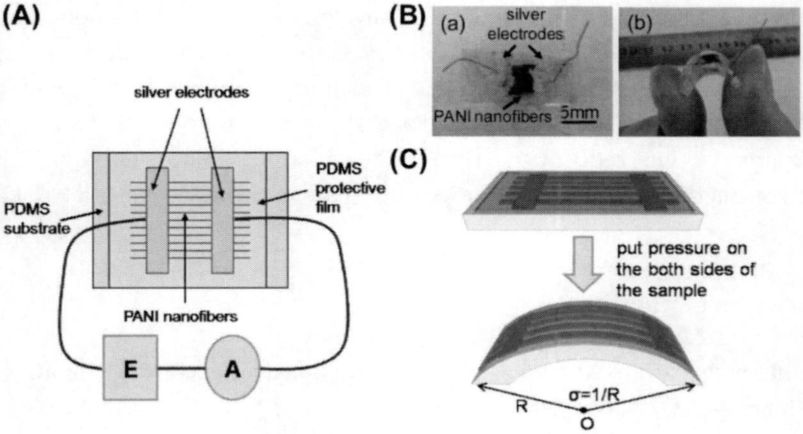

Figure 10. The composition of the nano-branched coaxial polyaniline / polyvinylidene fluoride fibers based strain sensors. (A) Diagrammatic sketch of the sample and the electronic test equipments. (a and b) in the picture (B) are the photograph of the sensor before and after strain respectively. (C) Schematic diagram for the strain sensor.

3.2. Field-Effect Transistors

Field-effect transistors (FETs) based on electrospun nanofibers have been reported extensively. For example, a single nanofiber FET from electrospun P3HT exhibited a hole field-effect mobility of 0.03 $cm^2 V^{-1} s^{-1}$ in the saturation regime, and a current on/off ratio of 10^3 in the accumulation mode [90]. Recently, transistors based on aligned P3HT nanofibers were fabricated by two-fluid coaxial electrospinning technique using P3HT as core and PMMA as shell, followed by extraction of PMMA. It is found that the carrier mobility of the aligned fibers at a lower shell flow rate (such as 1.0 mL h^{-1}) could be dramatically improved to 1.92×10^{-1} $cm^2 V^{-1} s^{-1}$, with the on/off ratio of 4.45×10^4 possibly due to higher crystallinity and preferred orientation of P3HT with electrospun aligned nanofibers [91].

These two values are one order of magnitude of higher than those of the single P3HT fibers [90]. These days a new continuous production of electrospun regioregular P3HT fibers by supplying additional solvent through an outer nozzle in the coaxial setup was demonstrated [93].

The key to the success was restricting the precipitation of P3HT at the nozzle tip, therefore preventing the nozzle from being clogged. The field effect mobility of P3HT fibers was 0.017 $cm^2\ V^{-1}\ s^{-1}$ at the saturation regime, and the on/off ratio was 100. In addition, González and Pinto [94] reported a new field effect transistor fabricated on the basis of electrospun regioregular poly(3-hexylthiophene-2,5-diyl) fiber. The hole mobility of the device was calculated to be $4\times10^{-4}\ cm^2\ V^{-1}\ s^{-1}$ and the on/off ratio was ~7. The large surface to volume ratio of the fiber makes it susceptible to doping. A large OFF current due to exposure to air is seen to affect the device characteristics.

3.3. Supercapacitors

Supercapacitor is known as electric double-layer capacitors, exhibiting 20-200 times greater capacitance and energy density (80-100 F/g, 5 Wh/kg) than conventional capacitors. Comparing with lithium ion batteries, supercapacitors also have the advantages of high power output, stable cycle life and faster energy release at high rate.

Recently, organic CPs, metal oxides, graphene/carbon nanotubes and their composites have been considered as promising candidates to conventional carbon-based electrodes in supercapacitors. For example, electrospun PANI nanofibers (PANI-NFs) as electrode materials for supercapacitor web were superior to PANI powder prepared by in situ chemical oxidative [95].

The specific capacitance of PANI-NFs web in 1 M H_2SO4 was 267 F g^{-1} at current density of 0.35 A g^{-1}, much higher as compared with PANI powder (208 F g^{-1}).

Furthermore, PANI-NFs web demonstrated very stable and superior performance than its counterpart and delivered specific capacitance 230 F g^{-1} at 1000th cycle. Capacitance retention of PANI-NFs web (86%) was higher than that observed for PANI powder (48%) indicating the feasibility of electrospun PANI-NFs web as superior electrode materials for supercapacitors. In addition, Miao et al. [96] fabricated hollow PANI nanofibers with controllable wall thickness. It achieved capacitance of 601 F g^{-1} at 1 A g^{-1}.

This may be attributed to their hollow structure, thin wall thickness, and orderly pore passages, which can facilitate the ion diffusion and improve the utilization of the electroactive PANI during the charge-discharge processes.

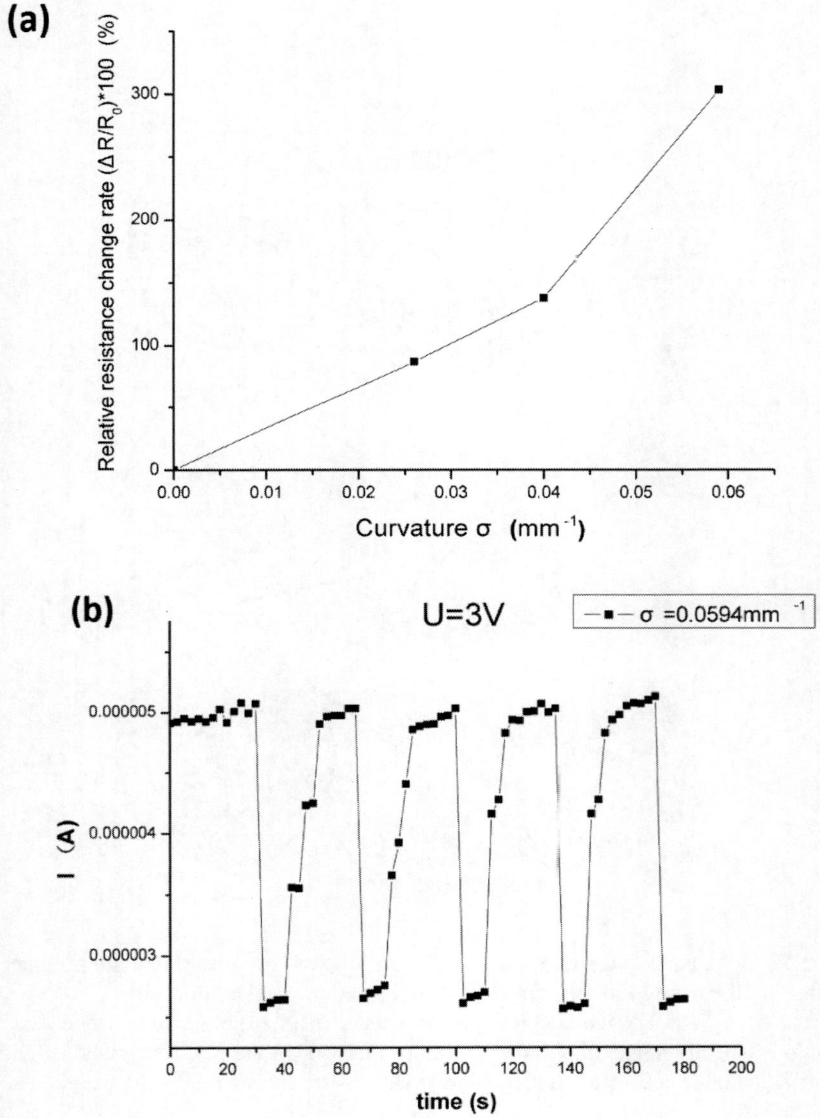

Figure 11. The strain sensitive property of the PANI/PVDF fibers web sample: (a) The strain sensitivity graph for this sample. (b) The sensitivity response graph of the repeated curving ($\sigma = 0.0594\ mm^{-1}$) and straightening experiment.

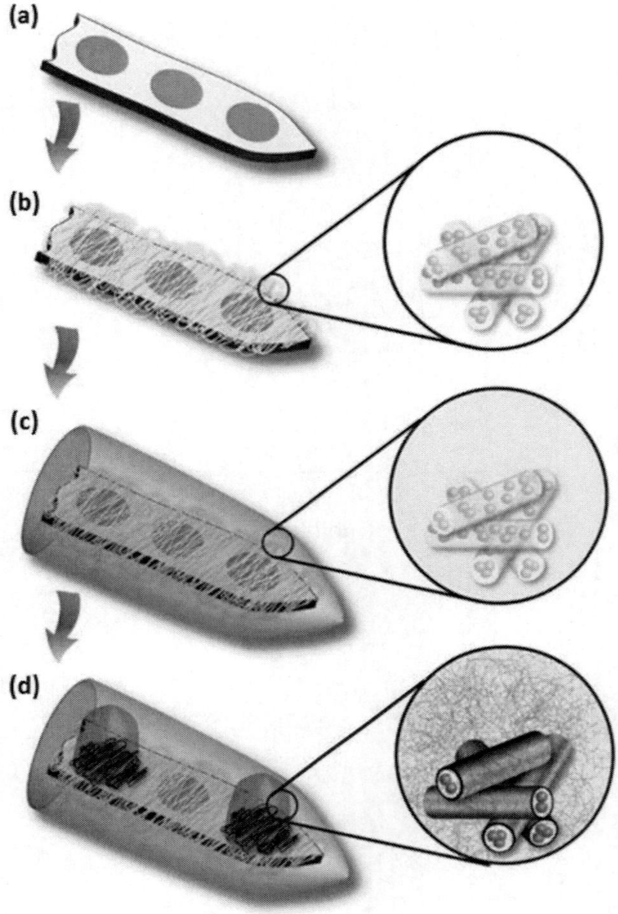

Figure 12. Schematic diagram of the fabrication process for multifunctional polymer coatings on the neural microelectrodes: (a) uncoated microelectrode, (b) electrospinning of DEX-loaded biodegradable nanofibers, (c) alginate hydrogel coating, (d) electrochemical polymerization of PEDOT on the electrode sites, around the DEX-loaded electrospun biodegradable nanofibers, and within the hydrogel scaffold. [101]

3.4. Neural Electrodes and Interfaces

Neural microelectrodes are capable of interfacing with the central and peripheral nervous system at the cellular level by transducing ionic biological signals to electronic signals. CPs such as PPY and PEDOT have become attractive candidates in the quest to bridge the electrode-cellular interface because they are compatible substrates for living cells. For example, PPY doped with p-toluene sulfonate is cyto-compatible with mouse fibroblasts and neuroblastoma cells, and it can support the regrowth of regenerating axons in vivo [96]. Particularly, it has been shown that PPY and PEDOT nanotubes can adhere better to the surface of electrodes in comparison with their film counterparts [97-100]. In this field, as an example, Abidian and Martin [101] designed a multifunctional, hybrid nanostructured interface for neural microelectrodes that is soft, low impedance, has high charge density, and is capable of controlled drug release. As shown in Figure 12, firstly, composite fibers of biodegradable poly(L, D-lactic-co-glycolic acid) (PLDL75G25A) and dexamethasone (drug) were fabricated by electrospinning and collected on the supporting microelectrodes (Figure 12b). Secondly, alginate was ionically crosslinked by Ca^{2+} ions using a dip coating method which covered the entire shank of microelectrodes (Figure 12c). Then PEDOT was electrochemically polymerized on the electrode sites, around the nanofibers and eventually inside the alginate hydrogel (Figure 12d). It is found that electrical properties of neural microelectrodes have been significantly improved with these coatings. The impedance of the neural microelectrodes can be significantly decreased by about two orders of magnitude (2.5±0.5 kΩ) and the charge capacity density significantly increased about two orders of magnitude (223.8±6.4 mC cm^{-2}) by creating PEDOT nanotubes on a gold electrode surface [97-100].

3.5. Other Electric Applications

Besides above mentioned applications, electrospun CP nanofibers also have other potential applications in electromagnetic interference shielding material [102] and electrically controlled color-changing textiles [103].

Conclusion

Owing to the unique properties and promising applications of CP nanofibers, there has been rapid growth in the development of CP ultrathin fibers via electrospinning in the past decade. This chapter briefly summarizes and reviews several different approaches of electrospinning (e.g., direct electrospinning of CPs into fibers, co-electrospinning of blends of CPs with other spinnable polymers, and template-assisted synthesis using electrospun fibers as templates) and several potential applications of CP nanofibers. However, in order to further enhance the performance of electrospun CP fibers-based devices and fulfill these interesting applications mentioned above, some major challenges remain to be confronted, such as further increase of solubility or spinnability of conventional CPs through modification and/or finding a better organic solvent or functional doping agent, further decrease of fibers' average outer diameter to within 100 nm, and further improvement of fibers' physical/chemical properties through electrospinning CP-based functional composite fibers.

ACKNOWLEDGMENTS

This work was supported by National Natural Science Foundation of China (51373082), Natural Science Foundation of Shandong Province, China for Distinguished Young Scholars (JQ201103), Taishan Scholars Program of Shandong Province (ts20120528), National Key Basic Research Development Program of China (973 special preliminary study plan, 2012CB722705), and Program for Scientific Research Innovation Team in Colleges and Universities of Shandong Province.

REFERENCES

[1] X. F. Lu, W. J. Zhang, C. Wang, T. C. Wen, Y. Wei, *Prog. Polym. Sci.*, 36 (5), 671-712 (2011).
[2] A. J. Heeger, *Angew. Chem. Int. Edit*, 40(14), 2591-2611 (2001).
[3] A. G. MacDiarmid, *Angew. Chem. Int. Edit*, 40(14), 2581-2590 (2001).
[4] C. M. Hangarter, N Chartuprayoon, S. C. Hernández, Y. Choa, N. V. Myung, *Nano Today*, 8(1), 39-55 (2013).
[5] T. A. Skotheim, J. R. Reynolds, Handbook of conducting polymers, 3rd edition. Boca Raton, CRC Press (2007).
[6] H. D. Tran, D. Li, R. B. Kaner, *Adv. Mater.* 21(14-15), 1487-1499 (2009).
[7] Y. Z. Long, M. M. Li, C. Z. Gu, M. X. Wan, J. L. Duvail, Z. W. Liu, Z. Y. Fan, *Prog. Polym Sci*, 36(10), 1415-1442 (2011).
[8] C. R. Martin, *Accounts Chem. Res.*, 28(2), 61-68 (1995).
[9] J. X. Huang, R. B. Kaner, *J. Am. Chem. Soc.*, 126(3), 851-855 (2004).

[10] M. X. Wan, *Adv. Mater.*, 20(15), 2926-2932 (2008).
[11] N. R. Chiou, A. J. Epstein, *Adv. Mater.*, 17(13), 1679-1683 (2005).
[12] J. Jang, H. Yoon, *Chem. Commun.*, 6, 720-721 (2003).
[13] Y. Huang, B. Dong, N. Lu, N. J. Yang, L. G. Gao, L. Tian, D. P. Qi, Q. Wu, L. F. Chi, *Small*, 5(5), 583-586 (2009).
[14] A. Formhals A. Process and apparatus for preparing artificial threads. US patent NO.1,975,504 (1934).
[15] D. H. Reneker, I. Chun, *Nanotechnology*, 7(3), 216-223 (1996).
[16] S. Megelski, J. S. Stephens, D. B. Chase, J. F. Rabolt, *Macromolecules*, 35(22), 8456-8466 (2002).
[17] R. Khajavi, M. Abbasipour, *Scientia. Iranica.*, 19(6), 2029-2034 (2012).
[18] T. T. T. Nguyen, C. Ghosh, S. G. Hwang, N. Chanunpanich, J. S. Park, *Int. J. Pharm.*, 439(1-2), 296-306 (2012).
[19] J. Bai, Y. X. Li, M. Y. Li, J. L. Gao, X. M. Zhang, S. G. Wang, C. Q. Zhang, Q. B. Yang, *Eng. Aspects*, 318(1-3), 259-262 (2008).
[20] S. Wang, Q. W. Liu, Y. Zhang, S. D. Wang, Y. X. Li, Q. B. Yang, Y. Song, *Appl. Surf. Sci.*, 279, 150-158 (2013).
[21] M. Eltohamy, U. S. Shin, J. E. Won, J. J. Kim, H. W. Kim, *Mater. Lett.*, 65(13), 2043-2046 (2011).
[22] J. F. Zheng, H. Y. Zhang, Z. G. Zhao, C. C. Han, *Polymer*, 53(2), 546-554 (2012).
[23] J. Tao, S. Shivkumar, *Mater. Lett.*, 61(11-12), 2325-2328 (2007).
[24] J. Yu, Y. J. Qiu, X. X. Zha, M. Yu, J. L. Yu, J. Rafique, J. Yin, *Eur. Polym. J.*, 44(9), 2838-2844 (2008).
[25] R. Kessick, G. Tepper, *Appl. Phys. Lett.*, 84(23), 4807-4809 (2004).
[26] X. C. Zhang, Y. P. Wang, D. B. Guo, Y. M. Wang, *AASRI Procedia*, 3, 236-241 (2012).
[27] K. L. Ou, C. S. Chen, L. H. Lin, J. C. Lu, Y. C. Shu, W. C. Tseng, J. C. Yang, S. Y. Lee, C. C. Chen, *Eur. Polym. J.*, 47(5), 882–892 (2011).
[28] H. N. Na, P. Chen, S. C. Wong, S. Hague, Q. Li, *Polymer*, 53(13), 2736-2743 (2012).
[29] Y. Zhao, X. Y. Cao, L. Jiang, *J. Am. Chem. Soc.*, 129(4), 764-765 (2007).
[30] K. Tang, Y. Yu, X. Mu, P. A. Van, Aken, J. Maier, *Electrochem. Commun.*, 28, 54-57 (2013).
[31] J. M. Lim, J. H. Moon, G. R. Yi, C. J. Heo, S. M. Yang, *Langmuir*, 22(8), 3445-3449 (2006).
[32] Y. Jin, D. Yang, D. Kang, X. Jiang, *Langmuir*, 26(2), 1186-1190 (2009).
[33] Y. Zhang, G. C. Rutledge, *Macromolecules*, 45(10), 4238-4246 (2012).

[34] R. Chen, C. Huang, Q. F. Ke, C. L. He, H. S. Wang, X. M. Mo, *Colloids and Surfaces B: Biointerfaces*, 79, 315-325 (2010).
[35] D. Li, Y. Wang, Y. Xia, *Nano Letters*, 3(8), 1167-1171 (2003).
[36] B. S. Jha, R. J. Colello, J. R. Bowman, S. A. Sell, K. D. Lee, J. W. Bigbee, G. L. Bowlin, W. N. Chow, B. E. Mathern, D. G. Simpson, *Acta Biomater*, 7(1), 203-215 (2011).
[37] J. S. Tan, Y. Z. Long, M. M. Li, *Chin. Phy. Lett.*, 25(8), 3067-3070 (2008).
[38] B. Sundaray, V. Subramanian, T. S. Natarajan, R. Z. Xiang, C. C. Chang, W. S. Fann, *Appl. Phys. Lett.*, 84(7), 1222-1224 (2004).
[39] M. K. Shin, Y. J. Kim, S. I. Kim, S. K. Kim, H. Lee, G. M. Spinks, S. J. Kim, *Sensors Actuators B*, 134(1), 122-126 (2008).
[40] Y. Z. Long, J. Zheng, F. Shao, W. H. Han, H. X. Yin, Z. H. Zhang, *China patent* No. CN201110137420.0 (2011).
[41] J. Rafique, J. Yu, J. L. Yu, G. Fang, K. W. Wong, Z. Zheng, H. C. Ong, W. M. Lau, *Appl. Phys. Lett.*, 91(6), 063126-063128 (2007).
[42] G. H. Kim, *Biomedical Materials*, 3(2), 025010-025018 (2008).
[43] M. Acharya, G. K. Arumugam, P. A. Heiden, *Macromolecular Materials and Engineering*, 293(8), 666-674 (2008).
[44] D. Yang, J. Zhang, J. Zhang, J. Nie, *J. Appl. Polym. Sci.*,110(6), 3368-3372 (2008).
[45] H. Wang, H. Tang, J. He, Q. Wang, *Mater. Res. Bull.*, 44(8), 1676-1680 (2009).
[46] A. Varesano, A. Montarsolo, C. Tonin, *Eur. Polym. J.*, 43(7), 2792-2798 (2007).
[47] E. Zhmayev, D. Cho, Y. L. Joo, *Polymer*, 51(18), 4140-4144 (2010).
[48] F. Dabirian, S. Sarkeshik, A. Kianiha, *Current Nanoscience*, 5(3), 318-323 (2009).
[49] M. M. Li, Y. Z. Long, D. Y. Yang, J. S. Sun, H. X. Yin, Z. L. Zhao, W. H. Kong, X. Y. Jiang, Z. Y. Fan, *J. Mater. Chem.*, 21(35), 13159-13162 (2011).
[50] D. Zhang, J. Chang, *Adv. Mater.*, 19(21), 3664-3667 (2007).
[51] N. J. Pinto, P. Carrión, J. X. Quiones, *Mater. Sci. Eng. A*, 366(1), 1-5 (2004).
[52] T. S. Kang, S. W. Lee, J. Joo, J. Y. Lee, *Synth Met*, 153(1-3), 61-64 (2005).
[53] Q. Z. Yu, M. M. Shi, M. Deng, M. Wang, H. Z. Chen, *Mater. Sci. Eng. B*, 150(1), 70-76 (2008).

[54] S. S. Srinivasan, R. Ratnadurai, M. U. Niemann, A. R. Phani, D. Y. Goswami, E. K. Stefanakos, *Int. J. Hydrogen Energy*, 35(1), 225-230 (2010).
[55] V. J. Babu, D. V. B. Murthy, V. Subramanian, V. R. K. Murthy, T. S. Natarajan, S. Ramakrishna, *J. Appl. Phys.*, 109(7), 0743061- 0743064 (2011).
[56] H. Bagheri, A. Aghakhani, *Anal. Chim. Acta.*, 713, 63-69 (2012).
[57] Q. Q. Lin, Y. Li, M. J. Yang, *Sens. Actuators B*, 161, 967-972 (2012).
[58] C. C. Tang, R. Huang, Y. Z. Long, B. Sun, H. D. Zhang, C. Z. Gu, W. X. Wang, J. J. Li, *Adv. Mater. Research*, 562, 308-311 (2012).
[59] R. Huang, Y. Z. Long, C. C. Tang, *Adv. Mater. Research*, 853,79-82 (2014).
[60] Q. F. Zhang, J. H. Wu, C. Whirang, M. K. Leach, E. W. Franz, Y. I. Naim, Z. Z. Gu, J. M. Corey, D. C. Martin, *Polymer*, 54(2), 702-708 (2013).
[61] D. Li, A. Babel, S. A. Jenekhe, Y. Xia, *Adv. Mater.*, 16(22), 2062-2066 (2004).
[62] W. Wang, Z. Y. Li, X. R. Xu, B. Dong, H. N. Zhang, Z. J. Wang, C. Wang, R. H. Baughman, S. L. Fang, *Small*, 7(5), 597–600 (2011).
[63] A. Huczko, *Appl. Phys. A*, 70(4), 365-376 (2000).
[64] S. Z. Ji, Y. Li, M. J. Yang, *Sens. Actat B*, 133(2), 644-649 (2008).
[65] Y. Zhu, J. C. Zhang, Y. M. Zheng, J. Zhai, L. Jiang, *Chem. J. Chin. Univ.*, 1, 196-198 (2006).
[66] Y. Zhu, L. Feng, F. Xia, *Macromol. Rapid Commun.*, 28(10), 1135-1141 (2007).
[67] E. Zampetti, S. Pantalei, A. Bearzotti, B. Bongiorno, F. D. Cesare, C. Spinella, A. Macagnano, *Procedia Engineering*, 47, 937-940 (2012).
[68] Y. H. Li, J. Gong, G. H. He, Y. L. Deng, *Synthetic Metals*, 161(1), 56-61 (2011).
[69] H. Bai, L. Zhao, C. H. Lu, C. Li, G. Q. Shi, *Polymer*, 50(14), 3292-3301 (2009).
[70] S. Nair, S. Natarajan, S. H. Kim, *Macromol. Rapid Commun.*, 26(20), 1599-1603 (2005).
[71] H. X. Wang, J. Ding, B. Lee, X. G. Wang, T. Lin, *J. Membr. Sci.*, 303(1), 119-125 (2007).
[72] F. Granato, A. Bianco, C. Bertarelli, G. Zerbi, *Macromol Rapid Commun*, 30(6), 453-458 (2009).
[73] S. Nair, E. Hsiao, S. H. Kim, *J. Mater. Chem.*, 18(42), 5155-5161 (2008).

[74] S. Nair, E. Hsiao, S. H. Kim, *Chem. Mater.*, 21(1), 115-121 (2009).
[75] A. Laforgue, L. Robitaille, *Macromolecules*, 43(9), 4194-4200 (2010).
[76] S. K. Li, X. F. Lu, X. Li, Y. P. Xue, C. C. Zhang, J. Y. Lei, C. Wang, *J. Colloid Interface Sci.*, 378(1), 30-35 (2012).
[77] S. H. Ku, S. H. Lee, C. B. Park, *Biomaterials*, 33(26), 6098-6104 (2012).
[78] A. Attout, S. Yunus, P. Bertrand, *Polym. Eng. Sci.*, 48(9), 1661-1666 (2008).
[79] K. Z. Yin, L. F. Zhang, C. L. Lai, L. L. Zhong, *J. Mater. Chem.*, 21(2), 444-448 (2011).
[80] B. Sun, Y. Z. Long, S. L. Liu, Y. Y. Huang, J. Ma, H. D. Zhang, G. Z. Shen, S. Xu, *Nanoscale*, 5(15), 7041-7045 (2013).
[81] H. C. Chang, C. L. Liu, W. C. Chen, *Adv. Funct. Mater.*, 23(39), 4960-4968 (2013).
[82] M. K. Tiwari, A. L. Yarin, C. M. Megaridis, *J. Appl. Phys.*, 103(4), 044305-044310 (2008).
[83] N. Liu, G. Fang, J. Wan, H. Zhou, H. Long, X. Zhao, *J. Mater. Chem.*, 21(47), 18962-18966 (2011).
[84] L. Persano, C. D. agdeviren, Y. Su, Y. Zhang, S. Girardo, D. Pisignano, Y. Huang, J. A. Rogers, *Nat. Commun.*, 4, 1633, DOI: 10.1038/ncomms2639 (2013).
[85] Q. Gao, H. Meguro, S. Okamoto, M. Kimura, *Langmuir*, 28(51), 17593-17596 (2012).
[86] M. Park, J. Im, M. Shin, Y. Min, J. Park, H. Cho, S. Park, M. B. Shim, S. Jeon, D. Y. Chung, J. Bae, J. Park, U. Jeong, K. Kim, *Nat. Nanotechnol.*, 7(12), 803-809 (2012).
[87] X. Chen, S. Guo, J. Li, G. Zhang, M. Lu, Y. Shi, *Sens. Actuators A*, 199(1), 372-378 (2013).
[88] L. Zhang, X. Wang, Y. Zhao, Z. Zhu, H. Fong, *Mater. Lett.*, 68(12), 133-136 (2012).
[89] J. Sirohi, I. Chopra, J. I. Mater, *Syst. Struct.*, 11(4), 246-257 (2000).
[90] H. Q. Liu, C. H. Reccius, H. G. Craighead, *Appl. Phys. Lett.*, 87(25), 2531061 (2005).
[91] J. Y. Chen, C. C. Kuo, C. S. Lai, W. C. Chen, H. L. Chen, *Macromolecules*, 44(8), 2883-2892 (2011).
[92] L. Sungwon, D. M. Geon, J. Unyong, *J. Mater. Chem.*, 19(6), 743-748 (2009).
[93] R. González, N. J. Pinto, *Synth. Met.*, 151(3), 275-278 (2005).
[94] S. Chaudhari, Y. Sharma, P. S. Archana, *J. Appl. Polym. Sci.*, 129(4), 1660-1668 (2012).

[95] Y. E. Miao, W. Fan, D. Chen, *ACS Appl. Mater. Interf.*, 5(10), 4423-4428 (2013).
[96] R. L. Williams, P. J. Doherty, *J. Mater. Sci: Mater. Med.*, 5(6-7), 429-433 (1994).
[97] M. R. Abidian, D. H. Kim, D. C. Martin, *Adv. Mater.*, 18(4), 405-409 (2006).
[98] M. R. Abidian, D. C. Martin, *Biomaterials*, 29(9), 1273-1283 (2008).
[99] M. R. Abidian, K. A. Ludwig, T. C. Marzullo, D. C. Martin, D. R. Kipke, *Adv. Mater.*, 21(37), 3764-3770 (2009).
[100] M. R. Abidian, J. M. Corey, D. R. Kipke, D. C. Martin, *Small*, 6(3), 421-429 (2010).
[101] M. R. Abidian, D. C. Martin, *Adv. Funct. Mater.*, 19(4), 573-585 (2009).
[102] J. S. Im, J. G. Kim, S. H. Lee, Y. S. Lee, *Physicochem. Eng. Aspects*, 364(1), 151-157, (2010).
[103] A. Laforgue, *J. Mater. Chem.*, 20(38), 8233-8235 (2010).

In: Advances in Conducting Polymers Research ISBN: 978-1-63463-258-4
Editor: Laura Michaelson

Chapter 4

CHARGE TRANSFER AND ELECTROCHEMICAL REACTIONS AT ELECTRODES MODIFIED WITH PRISTINE AND METAL-CONTAINING FILMS OF CONDUCTING POLYMERS

V. V. Kondratiev[1], O. V. Levin[1] and V. V. Malev[1,2]
[1]Institute of Chemistry, St. Petersburg State University (SPbU), St. Petersburg, Petrodvoretz
[2]Institute of Cytology, Russian Academy of Sciences, St. Petersburg, Russian Federation

ABSTRACT

The review is based mainly on the experimental results obtained with electrode systems consisting of different substrates modified by such typical conducting polymers, as polythiophenes and nickel polymer complexes with the Schiff base ligands. The established electrochemical properties of these modified electrodes, as well as the obtained data of their spectroelectrochemical and quartz crystal microbalance studies are discussed in the main part of the review. The performed comparison between these results and those followed from the accepted theory of charge transfer in modified electrodes shows only their qualitative agreement, so that the necessity of improving the existing representations becomes evident. Different methods of syntheses of metal-containing films based on conducting polymers are shortly discussed in connection with the subsequent studies of some electrochemical processes occurring at such composite electrodes. A new approach to treating the polaron conductance of polymer films is proposed. As shown, its inferences significantly differ from the predictions of the existing theory. This permits one to consider the proposed approach as some premise for more detailed studies.

1. INTRODUCTION

Since discovery of a new class of materials usually called as “conducting polymers” (CPs) in 1977 [1], a lot of new research directions with a use of these materials has been developed. They include not only studies of different CPs and their copolymers, but also a great variety of new composite materials based on CPs with inclusions of metal or metal oxide particles, other inorganic compounds etc. Such development was stimulated by practical applications of these materials in the fields of electrocatalysis, energy storage, electrochemical sensors and biosensors, molecular electronics, photoelectrochemistry, electroluminescence and electrochromic devices, antistatic and anticorrosion coatings, microwave and radar-adsorbing materials [2-14].

Conducting polymers can be obtained from molecular monomers containing aromatic benzoid or non-benzoid amines (like aniline, *o*-aminophenoles, *o*-phenylendiamine), heterocyclic molecules and its derivatives (pyrroles, thiophenes, carbazoles etc.), which, under their oxidation, form conjugated polymer chains with alternated single and double bonds. The most important groups of CPs possessing the high electronic conductivity are polyaniline family (PANI), polythiophenes (PTh), and polypyrroles (PPy). CPs belong to a class of substances with a mixed electronic–ionic conductance and their conductive properties change between quasi-metallic and semiconductor ones. During the initial period of CP studies, the main research interests have been focused on syntheses of new polymers and understanding of the mechanisms of charge transport into CPs. Nowadays, the major research interest has shifted to studying new materials based on CPs, their nanostructuring, and producing different composite films that include metal and metal oxides nanostructures [15-25].

Electrochemical methods have played an important role in elaboration of practical procedures of syntheses of CPs and composite films, their characterization, and different applications. This results from the high conductivity of polymers and the evident possibility to control the key electrochemical parameters (current, potential, charge) during the polymer syntheses, which allows one to vary the morphology and properties of the obtained materials. However, chemical routes of an oxidative polymerization are also widely used for preparations of conducting polymers.

A broad spectrum of problems related to CPs and CP-based materials investigations, as well as a great number of methods used in such studies do not allow to describe completely all aspects of the research in the field under discussion in scope of a separate review article. Therefore in this work, we will restrict ourselves only to consideration of electrochemical (and related) aspects of such research and illustrations of using some models at treatment of electrode processes occurring in a series of particular systems.

Polythiophenes (PTh) and polianilines (PANI) belong to the most intensively studied families of CPs, because of simple procedures of their syntheses, their stability, the high conductivity, and possible applications in different devices [2-5,11-14]. Among CPs with conjugated bonds, PTh films exhibit a relatively simple electrochemical behavior due to the absence of protonation effects, which are characteristic for PANI films in aqueous solutions. Electrodes modified with polymer complexes of transient metals with the Schiff bases are also the permanent subjects of intensive studies due to the same reasons. Accounting for these aspects and owing large experience obtained during the corresponding studies, we will discuss below the main patterns of CP electrochemical behavior with aforementioned systems.

2. Electrochemical Properties of Conducting Polymers

2.1. Conductivity of CPs. Polaron-Bipolaron Concept

A fundamental breakthrough in CP studies was made with introduction of the polaron-bipolaron concept of conductivity, explaining the mechanism of charge transport into CP-modified electrodes [3-6, 14, 26-29]. Now it is well recognized that the electrochemical behavior of CPs and their conductivity are based on the electrode processes known as *p*- and *n*-doping ones, which, in electrochemical terms, corresponds to oxidation and reduction of the polymers with switching between quasi-metallic and insulating states. The doping/dedoping processes in CPs are related to the mixed electronic and ionic transport in the bulk of such materials. Formation of positively charged delocalized states (polarons) during a *p*-doping process is accompanied with corresponding insertion of charge-compensating species, which in the most cases are doping anions (in general case, counter-ions) [3-6]. Translocations of delocalized electrons or holes along polymer chains obviously lead to the accompanying movement of counter-ions. However, if polyanionic species are used as supporting electrolyte during CPs syntheses, for instance PSS–anion, such anion-dopant is immobile and other more mobile ions act as charge-compensating ions [11, 12].

As mentioned above, the general polaron concept accepted in the physics of condensed matter was expanded to the case of organic conducting polymers. An electrochemically active fragment of CPs is a double bond. Electron of a π-bond in a π-conjugated polymer can be easily removed or added electrochemically to form a polymeric ion without much disruption of σ–bonds, which corresponds to *p*- or *n*-doping of the polymer. The charge formation promotes a local distortion in the polymer backbone and the arising charged quasi-particles together with their local surroundings are called as polarons or bipolarons [26-29].

EPR studies of CP films confirm the presence of polarons (radical cations) with unpaired electrons in such systems [30-31]. The formed radical cation is spatially delocalized over a number of polymer chain fragments (usually 3÷6 monomer units). This leads to deformation of the polymer structure and polarization of the environment (this gave the name `polaron' to the chain fragments containing a radical cation). Polarization of the environment and interactions with both the neighboring polymer chains and counter-ions lead to stabilization of such radical cations. The polaron state might undergo further oxidation with formation of bipolaron, *i.e.* a pair of positively charges of opposite spins (dication). At present, the polaron theory of conductivity has been accepted as the main model for explanation of the electronic conductivity of polymers with conjugated bonds. Theoretical description of charge transfer processes in conducting polymers is based on this polaron-bipolaron concept and usually explores the model of "hopping" mechanism, *i.e.* the charge exchange between two neighboring quasi-particles of different oxidation states [32-40].

2.2. Cyclic Voltammetry of CP films

CPs can be prepared as relatively thick films on the electrode surfaces (conducting substrates) by electrochemical oxidation of monomers. A cyclic voltammetric (CV) response of CP-modified electrodes in electrolyte solutions usually shows a broad, not well-expressed cathodic and anodic peaks with a pronounced current plateau [6, 43, 44].

Figure 2.1a,b shows the typical CV curves of some polythiophenes with different substituents in the thiophene rings: namely, poly-3-methylthiophene (PMT) [43-49], poly-3-octylthiophene (POT) [50-54], poly-3-dodecylthiophene (PDDT) [41, 50], poly-3,4-ethylenedioxythiophene (PEDOT) [55-58]. More or lesser expressed current peaks are observed for POT and PDDT films, whereas, in the case of PMT and especially PEDOT films, CV curves display very broad peaks in a wide range of electrode potentials.

The potential ranges of CVs for a series of polymers given in Figure 2.1 differ essentially from each other due to some differences in the positive limits of applied potentials, where the polymers are stable. It is also worth to note that PEDOT films having the cyclic electron-donated ethylene substituent in thiophene rings display a high electroactivitity in a very wide potential range at using both aqueous and non-aqueous solutions.

Due to its electronic structure, PEDOT is one of the few conducting polymers that can be both *n*- and *p*-doped polymer, and its behavior significantly differs from other polythiophene derivatives [11, 12].

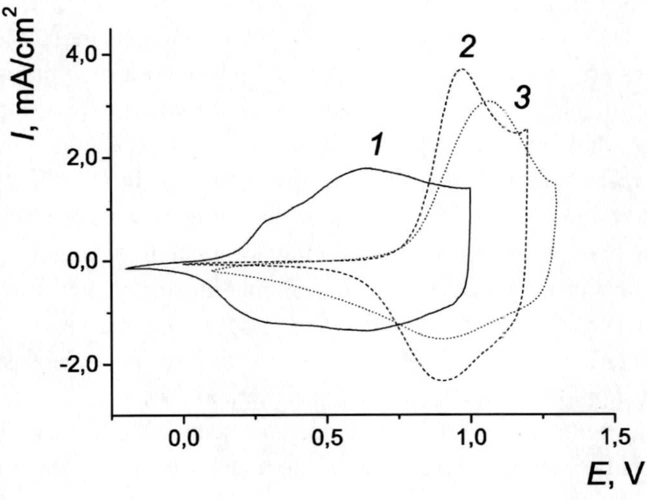

Figure 2.1a. Cyclic voltammograms (CVs) of PMT (1), POT (2), PDDT (3) films in *0.5 M* $LiClO_4$ acetonitrile solution (scan rate 50 mV s^{-1}).

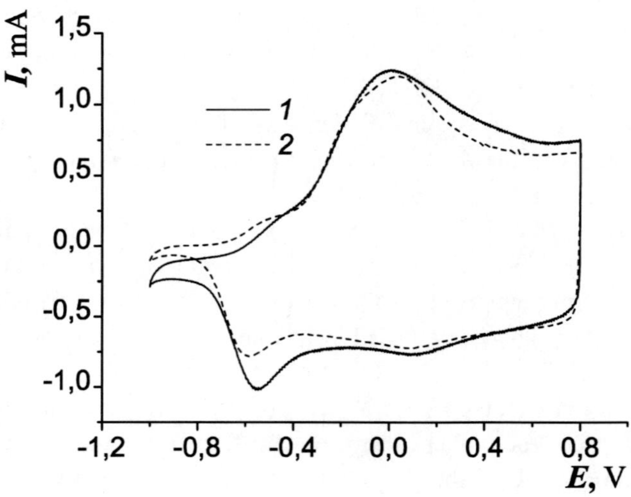

Figure 2.1b. Cyclic voltammograms of PEDOT film in *0.5 M* $LiClO_4$ acetonitrile (1) and propylene carbonate (2) solutions (scan rate 50 mV s^{-1}).

For all polythiophene-modified electrodes, current responses are proportional to the rate of sweeping the potential, which indicates to a quasi-equilibrium character of CV-curves, *i.e.* the absence of diffusion obstacles within the film interior. Voltammetric behavior of the most CPs is characterized with stable current responses after first two-three cycles and these responses are proportional to the charge consumed at polymerization of the film tested. In case of PEDOT films, such proportionality takes place up to polymerization charge values equal to 60 C cm^{-2} (with the film thickness of about 0.5 mm), as reported in [59]. The appearance of broad CV peaks has been explained by the existence of similar redox systems with different formal potentials (due to some differences in a conjugation length of separate polymer chains, their local environment, structural inhomogeneity of the films [6], and the presence of negative interactions between the forming charges).

Cyclic voltammetry is probably the most widely used electrochemical method, but it should be considered only as a method for a primary studying of CPs due to a multistep character of the electrode processes within such systems. Actually, the information to be obtained from such measurements reduces only to the peak current and potential values, as well as their dependences on the rate of sweeping the electrode potential.

Many attempts have been carried out on earlier stages of studying CP-modified electrodes in order to evaluate diffusion coefficients of charge carriers within the films investigated. However, the values reported in the literature seem often to be unreliable, since they vary in their magnitude by several orders according to different works. Such scattering is, most probably, a consequence of some structural variety of CPs obtained in different works. At the same time, it might also result from some unreliability of the calculations performed. As shown in [25], implementation of the Randles-Shevchik equation in case of modified electrodes requires adequate redefinitions of its parameters, which have more complex meanings for such systems than those assumed according to the original equation. Here, one more possible source of some errors should also be indicated.

During a relatively long period, it has also seemed that some important conclusions might be extracted from such characteristics of CV curves as their mid-width. These expectations were based on the representations developed earlier by Laviron in case of reactions proceeding in compact monolayer films of electroactive particles adsorbed onto metal electrodes [60,61]. However, the criteria proposed by Laviron to establish the presence and character of the so-called short-range interactions within these films are only qualitatively valid for multilayer ones, as shown in [62]. This results from the fact that modifying films of CPs, as a rule, have a phase character in contrast to monolayers of adsorbed particles and, hence, the first ones are electroneutral formations. The latter means that the concentrations of charge carriers inside such films depend on the electric potential value within their interior and this circumstance is not accounted for in scope of the Laviron's consideration (see review [63]).

The above said determines the reasons why, in parallel to CV measurements, application of more sophisticated techniques and combined methods is necessary for detailed studies of charge transport in CPs.

2.3. Electrochemical Impedance Spectroscopy

Electrochemical impedance spectroscopy (EIS) is recognized as one of the most powerful technique for evaluations of the parameters of charge transfer and diffusion processes in polymer-modified electrodes immersed into electrolyte solutions [33, 64]. Among CPs with conjugated bonds PPy and PTh films exhibit a relatively simple electrochemical behavior. When using non-aqueous solutions these systems are free from the problems resulting from protonation effects inherent to PANI films, for example.

The schematic impedance spectra presented in Figure 2.2 show typical features corresponding to the parallel resistance (R) - capacity (C) junction, the existence of the Warburg impedance, and the capacitive-like behavior [32, 33]. A similar EIS pattern with the indicated characteristic parts of its spectrum was found for a series of polymer modified electrodes. For instance, a suppressed high-frequency semicircle and an extended (with respect to the frequencies interval) capacitive response were observed for polypyrrole (and its derivatives) films on various substrates in propylene carbonate solutions of $LiClO_4$ [65-69] and aqueous KCl solutions [70].

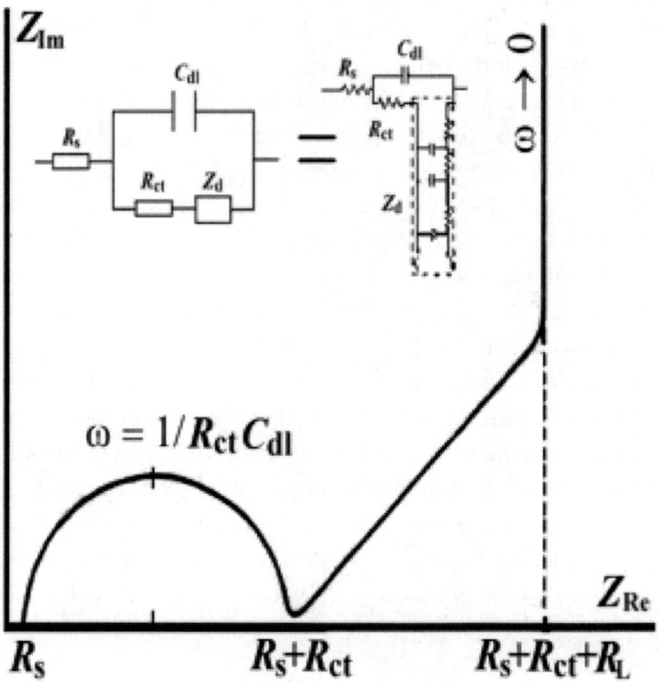

Figure 2.2. Schematic impedance spectra, corresponding to the Rendles circuit (inset).

The same parts in the impedance spectra were registered for polythiophene derivatives, namely, poly(3-methylthiophene) (PMT) [43-49], poly(3-octylthiophene) (POT) [50-54], poly(3,4-ethylene-dioxythiophene) (PEDOT) [55-58], and some other compounds [71,72]. It was also reported that the impedance spectra of the reduced or weakly oxidized forms of polythiophenes exhibit not only the high-frequency semicircle regions and low-frequency capacitive responses, but also the parts corresponding to the Warburg impedance with their slopes approaching to unity [51, 52, 54]. The high-frequency semicircles were mostly pronounced for a slightly doped polypyrrole [66, 70]. The presence of such semicircles and noticeable capacitive responses were also reported for poly(3,6-dioxyheptylpyrrole) and poly(3-methylthiophene) films in propylene carbonate 1M $LiClO_4$ solutions [67]. However, it should be pointed out here that the above semicircles are sometimes distorted and correspond to negative values of Z_{Re}-axis in the limit of high frequency, which probably connected with some irregularities in working the reference electrodes at these frequencies. Thus, the impedance spectra of polypyrrole- and polythiophene-based polymer films exhibit characteristic parts with well-separated frequency regions that correspond to the limiting influence of different stages of charge transfer within these objects. Below, we will represent the results obtained for particular systems, in which the impedance parameters were systematically estimated.

In scope of the existing theory of modified electrodes [32-38], no principal differences are present in a quantitative treatment of the redox and polaron conductance. The only thing accounted for in case of polaron-containing films is the established fact that a polaron quasi-particle includes several monomer units (3÷6, as pointed out above) of the polymer chain, along which the polaron moves. However, the subsequent use of the same theoretical equations (in particular, the Mathias & Haas equations of the modified electrode impedance [32]) as those derived, in essence, for redox-polymer films eliminates the above peculiarity of polaron quasi-particles. Postponing a more detailed explanation of this statement up to the seventh section of the review, we would like to emphasize here that the most part of diffusion coefficient determinations (including those given below) are based on the indicated theory. This means that the parameters obtained from the corresponding fitting might be considered as reliable ones only if a complete agreement between the used theory and experiment is observed. As will be clear from the further discussion, such complete correspondence does not take place usually.

Typical impedance spectra exhibiting three noticeable regions (quasi-capacitive, Warburg, and high-frequency ones) were observed for polypyrrole films in aqueous electrolyte solutions [66]. Using these data, the electron diffusion coefficient for polypyrrole films was estimated as D_e~10^{-3} cm^2 s^{-1}. It was also found that the ion diffusion coefficient D_m for the film (L=15 μm) strongly depends on the anion nature of the electrolyte used, namely, D_m=$2.56\cdot10^{-6}$ (Cl^-), $6.2\cdot10^{-7}$ (ClO_4^-), $8.2\cdot10^{-8}$ (SO_4^{2-}), and $1.6\cdot10^{-9}$ cm^2 s^{-1} (tosylate ion), but is weakly dependent on the cation nature (H^+, Na^+, Me_4N^+).

The impedance spectra of non-substituted polythiophene films were analyzed [73] in functions of the electrode potential and the film thickness for an asymmetric configuration of the metal/polymer/solution system (that is characteristic for modified electrodes) and a symmetric one (namely, metal/polymer/metal with using an air-dry film). The high-frequency semicircle in the impedance spectrum of the asymmetric configuration and the derived parameters of a model RC-circuit (R_{ct} =30 Ohm·cm^2 and C_{dl} =1÷5 μF·cm^{-2}) were attributed to the film-solution interface. However experimental indications to such localization of the chosen interface (substrate or solution) were, strictly speaking, absent.

The platinum electrode modified with a PMT film was studied in a 0.2M $LiClO_4$ solution in MeCN by EIS method [43]. As established, the impedance spectra of the reduced PMT form exhibit high-frequency semicircles, which are gradually transformed into quasi-capacitive responses if the applied frequency decreases. When going to the oxidized form of PMT, the diameter of such semicircles significantly decreases. The diffuse part (*i.e.* the Warburg one) of the spectr was not practically registered in these measurements.

Figure 2.3 shows the typical impedance spectra of PMT, POT, PDDT, and PEDOT obtained in acetonitrile solutions of 0.5M $LiClO_4$. The results presented in Figure 2.3 were partly published in our papers [49,54,74,75]. Strongly suppressed or partly pronounced high-frequency semicircles were observed for electrodes modified with PEDOT and PMT films. This is in agreement with the impedance spectra obtained for similar conditions in a series of other works [55, 58] for PEDOT). Analogous shape of the impedance spectra of PMT films was registered in a 0.1M $NBun_4PF_6$ solution in propylene carbonate [45]. The mid-frequency parts of the spectra for PMT and PEDOT did not exhibit a well expressed pattern of the Warburg impedance and quickly transformed into abruptly rising curves associated with the quasi-capacitive response of the films. The charge transfer resistance (30-60 Ohm cm^2) determined from the semicircles that observed for PMT was attributed to the film/electrolyte solution interface [45, 49].

Results of the impedance studies of PMT films deposited on platinum in aqueous electrolyte solutions containing different anions were also reported [46]. It was concluded that the charge transfer resistance at the film/solution interface contributed largely to the measured impedance. The overall patterns of the impedance spectra of PMT films were also reported in work [50], however its authors drew only qualitative conclusions on the appearance of incomplete high-frequency semicircles and low-frequency pseudo-capacitive responses.

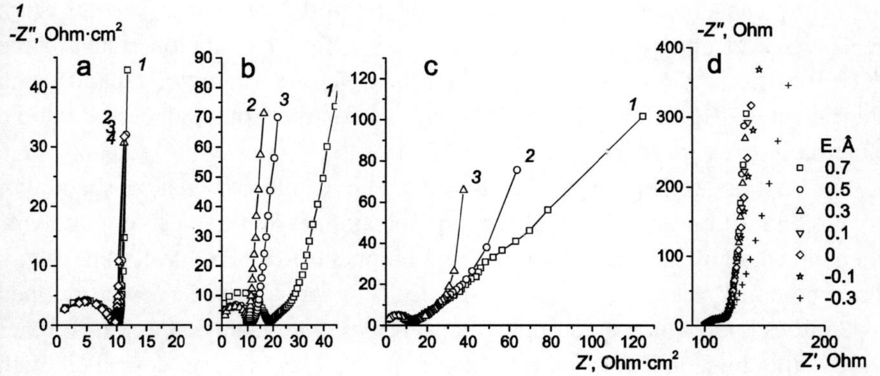

Figure 2.3 a,b,c,d. Impedance spectra of PMT (a), POT (b), PDDT (c) and PEDOT (d) films in 0.5 M $LiClO_4$ acetonitrile solution at different potentials E,V: *a: –1* – 0.5, *2* – 0.7, *3* – 0.9, *4* – 1.0; *b :1* – 0.8, *2* – 0.9, *3* – 1.1; *c: 1* – 0.8, *2* – 0.9, *3* – 1.0; d: variation in the range -0,3 ÷ 0.7.

Significantly different impedance spectra have been obtained for the neutral and oxidized forms of POT and PDDT films in acetonitrile solutions of lithium perchlorate [49, 54] (see Figure 2.2b). In addition to the characteristic high-frequency semicircle and pseudo-capacitive response, the transient Warburg type region is appeared in these spectra. The diffusion impedance components become well-pronounced ones, when analyzing the impedance data in the $-Z_{Im}$, Z_{Re} vs. $f^{-0.5}$ coordinates. This analysis reveals parallel linear dependencies indicating diffusion-controlled charge transfer in the film interior.

The diffusion coefficients occurred be equal to 2.9·10-8 cm2·s-1 (E = 0.8 V), 6.8·10-8 cm2·s-1 (E = 0.9 V), 1.5·10-7 cm2·s-1 (E = 1.0 V), 6.3·10-7 cm2·s-1 (E = 1.1 V), 1.7·10-6 cm2·s-1 (E = 1.2 V), that is, they were dependent on the electrode potential.

Similar results were obtained in [51,53] for Pt/POT-electrode immersed in a 0.1M solution of $LiBF_4$ in propylene carbonate. The spectra of the neutral form of POT exhibited a semicircle corresponding to a rather high resistance R_{ct} (about 30 kOhm), which was explained by the effects of the double layer on the polymer/ electrolyte solution interface with C_{dl}=2 $\mu F \cdot cm^{-2}$. For the doped (oxidized) form of POT the semicircle becomes more pronounced and its diameter considerably decreases as the degree of film oxidation increases, namely, R_{ct} decreases to 980-790 Ohm at the electrode potential equal to 1.2 V. At the potential changes in the range 0.8÷1.2 V, the capacitance C_{dl} increased by nearly a factor of 1.5 (from 2.8 to 4 $\mu F \cdot cm^{-2}$) and the low-frequency capacitance C_{lf} changed from 0.76 to 4.1 mF·cm^{-2}. The diffusion coefficients of charge carriers determined at different oxidation states of the POT film lie in the range (1÷8)·10^{-8} cm^2·s^{-1}. Similar values of the diffusion coefficient of BF_4^- ions (D_{eff}=8 10^{-9} cm^2 s^{-1} for the potential of the anodic maximum in CV) were reported for POT films [51].

In contrast to the spectra considered above, the impedance spectra of polyaniline films and films based on polyaniline derivatives, e.g. poly(o-phenylenediamine) (PPD) and poly(o-aminophenol) (PAP), have, as a rule, a more complex shape. This manifests itself in the absence of well-defined spectral regions, which could be associated with particular processes. In other words, the impedance spectra of these systems cannot be described with simple theoretical equations. The impedance of polyaniline films was also studied in hydrochloric acid [76]. Dependences of the capacitance and conductivity on the electrode potential as measured ones at a constant frequency of 10 Hz exhibit maxima in the region of electrical activity of the film. The authors of work [77] performed a thorough analysis of the published data [76] and concluded that the properties of polyaniline films can be explained only with allowance for the porous film structure.

The EIS spectra of PPD films exhibit a region corresponding to the Warburg impedance and a poorly pronounced low-frequency capacitive response [78]. Analysis of the impedance spectra and estimation of the corresponding parameters were carried out in scope of the homogeneous film model [32]. It was established that the low-frequency capacitance is proportional to the film thickness and the maximum capacitance is achieved at the polymer formal potential. A minimum of the Warburg constant and the maximum width of the Warburg impedance region were also observed near the formal potential. The effective diffusion coefficients are of the order of (1÷10)·10^{-8} cm^2 s^{-1} depending on the electrode potential [78]. At the same time, it was reported that the used model of homogenous films is irrelevant to describe specific features of the processes occurring in the polymer, in particular, protonation of the film redox centers. The study of the impedance spectra of PPD film on a glassy carbon substrate in a broad range of pH values allowed the authors of [79] to analyze the dependences of charge transfer parameters on the hydrogen ions concentration. It was found that, at low pH values, the observed impedance spectra were slightly different from those published earlier [78, 80] and the high-frequency semicircles appeared in the spectra only at pH > 4.

Such key parameters, as interfacial charge transfer resistance R_{ct}, low-frequency capacitance C_{lf}, and effective diffusion coefficient D_{eff} were determined in [79] by using relationships of the homogeneous film impedance [32]. It was also established that capacitance C_{lf} and diffusion coefficient D_{eff} depend on the electrode potential and reach maximal values near the formal potential. The observed increase in D_{eff}-values with a decrease in pH was treated as an indication to the fact that electron transfer is accompanied by an intermolecular proton transfer.

The role of porosity in charging CP films has been discussed in the context of the `faradaic' and `capacitive' components of the overall current [81-89]. A number of methods for experimental separation of these contributions was proposed. Among them, the most important is a spectroelectrochemical research of CP films in UV and visible spectral regions [89-98]. However, the problem of interpreting the capacitive responses has not been solved so far in complete form.

When analyzing the experimentally determined low-frequency capacitance C_{lf} of electroactive polymer films in the region of their redox activity, it usually occurs to be a parabolic function of the electrode potential with a maximum at oxidation state ***y*** of about 0.5 (Figure 2.4). This results from the fact that the maximum value of the product of the concentrations of the oxidized and reduced fragments in the polymer corresponds to the formal potential. However, the observed dependences of C_{lf} on the electrode potential are only in a qualitative agreement with the theoretical equation for the low-frequency capacitance, which follows from the homogeneous film model

$$C_{LF} = \frac{AF^2L}{RT}\left\{\frac{1+a\theta_R^0(1-\theta_R^0)}{C_R^0(1-\theta_R^0)} + \frac{1}{z_m^2C_m^0}\right\}^{-1} \tag{2.1}$$

The $C_{lf}(E)$-dependence calculated for poly(3-octylthiophene films [54,74] according to Eqn (2.1) fits the experimental curve only if one assumes the interaction parameter value, ***a***(see eqn (2.1)), be equal to 6.8. According to the definition of the parameter, its obtained sign and magnitude indicate to high repulsive interactions between redox fragments in the film. The same character of the experimental $C_{lf}(E)$-curve was established for PPD films [78].

Possible reasons for significant differences between theoretical and experimental curves include structural inhomogeneity of CP films that leads to dispersion of the formal redox potentials and film swelling effects; both of them are ignored in the homogeneous film model [54, 74].

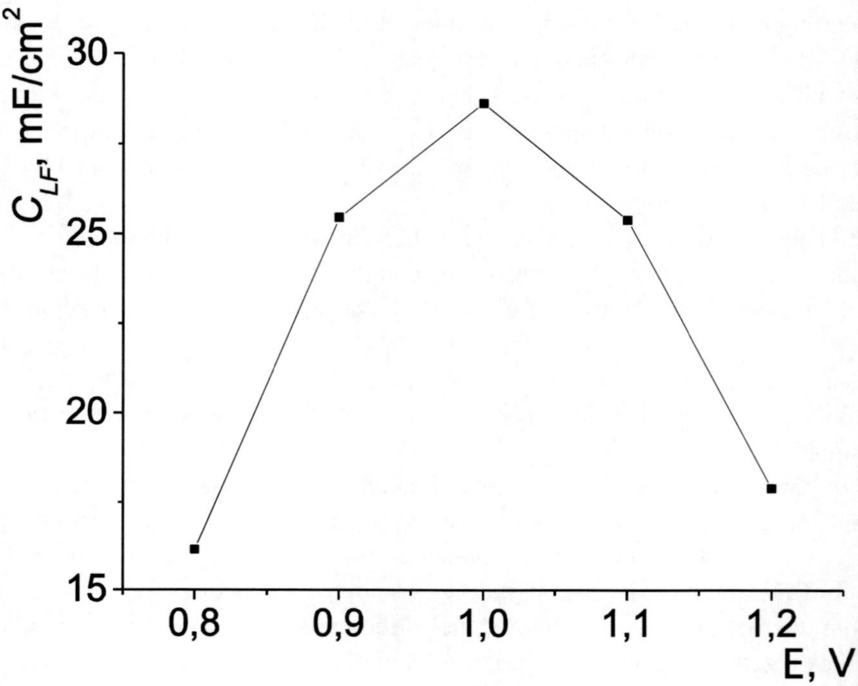

Figure 2.4. Dependence of low-frequency capacitance of POT film vs. electrode potential in 0.5 M $LiClO_4$ acetonitrile solution.

It seems that these factors affecting the low-frequency capacitance cannot be accounted for even formally with the help of parameter ***a***.

Analogous fitting of the experimental curve of C_{lf} vs. E was performed for PDDT film, where a satisfactory agreement between the results compared was found at interaction parameter ***a***=7.4. This value seems to be too high for repulsive interactions. However, one can consider the data obtained for POT and PDDT polymer films to be in a qualitative agreement with theoretical predictions.

Dispersion of the low-frequency capacitance is characteristic for the impedance spectra of CPs. Up to now the nature of such dependence is a debated topic. Considerable attention to this issue was paid by Bisquert et al. [68, 69, 99, 100], who reported results of a detailed analysis of the low-frequency capacity dependence on the thickness of polypyrrole films and the electrode potential. In particular, the frequency dispersion of the capacity was explained by the existence of film porosity and the experimental data were treated in scope of the modified Albery's model for charge transmission lines

[101-103]. However, from this analysis it follows that the specific low-frequency capacitance $C = C_{lf}/L = 100{\div}200\ F{\cdot}cm^{-3}$ is almost independent of the film thickness. This points to a linear dependence of C_{lf} on the film thickness and thus, suggests the volume character of the film charging rather than formation of charged double layers on the large inner surface of the porous film. In this case, the dispersion of capacity is associated with effects of a diffusion length (pore size and polymer grain size), spatial distribution of charge carriers over the bulk polymer, and binding of counter-ions in the film [68].

We have spared a sufficiently high attention to the above discussions of the cyclic voltammetry and impedance results obtained with the objects in question. This was stipulated by our wish to clearly demonstrate that, basing on such data, one can say only about some qualitative accordance between the existing theoretical and experimental results. Actually, in all the cases considered, it occurs necessary to draw additional arguments in order to explain the observed deviations (or abnormalities) of experimental results with respect to the theoretical predictions. Certainly, this might be the consequence of a real complexity of the studied systems, but such explanation is only a verbal one. On the other hand, the absence of a complete (quantitative) agreement between the compared results allows one to have some doubts as to validity (or sufficiency) of the applied theoretical representations for polaron-conducting polymer films.

References to Sections 1, 2

[1] H. Shirakawa, E. J. Louis, A. MacDiarmid, C. K. Chiang, A. J. Heeger, *J. Chem. Soc., Chem. Commun.*, (1977) 578.

[2] G. Tourillon, in *Handbook of Conducting Polymers* (Ed. T A Skotheim) (New York: Marcel Dekker, (1986) 293

[3] J. Heinze, Electronically Conducting Polymers, *Topics in Current Chemistry,* 152, Springer, 1990. P. 1.

[4] G. Schopf, G. Kossmehl *Polythiophenes - Electrically Conductive Polymers,* Berlin: Springer, 1995

[5] T.A. Skotheim, R.L. Elsebaumer, J. R. Reynolds, Handbook *of Conducting Polymers*. 2nd Ed. / Eds. N.Y.: Marcel Dekker, 1997. 1075 P.

[6] G. Inzelt,M. Pineri,J.W. Schultze, M.A. Vorotyntsev, *Electrochim.* Acta., 45 (2000) 2403.

[7] P. Novak, K. Muller, K. S. V. Santhsanam, O. Haas, *Chem. Rev.*, 97 (1997) 207.
[8] G. Inzelt, in *Electroanalytical Chemistry* Vol. 18 (Ed. A J Bard) (New York: Marcel Dekker, 1994) p. 89.
[9] R. W. Murray, in *Electroanalytical Chemistry* Vol.13 (Ed. A J Bard) New York: Marcel Dekker, 1984, p. 20
[10] P. Chandrasekhar, *Conducting Polymers. Fundamental and Applications*. Kluwer Academic Publishers. Boston. 1999, 645.
[11] L. Groenendaal, F.J onas, D. Freitag, H. Pielartzik, J. Reynolds R. *Adv. Mater.,* 12 (2000) 481.
[12] L. Groenendaal, G. Zotti, Aubert, P.-H. Waybright, S.M. Reynolds, J. R. *Adv. Mater.,*15 (2003) 855.
[13] J. Roncali, Conjugated polythiophenes: Synthesis, functionalization and applications. *Chem. Reviews*., 92 (1992) 711.
[14] G. Inzelt, *Conducting polymers: New era in electrochemistry*. Springer. (2008) 282.
[15] J. Wang Nanomaterial-based electrochemical sensors, *Analyst,* 130 (2005) 421.
[16] V. Tsakova Metallic-Based Nanocomposites of Conductive Polymers, 289. (In book "*Nanostructured Conducting Polymers*" Ed. by A. Eftekhani, A John Wiley and Sons, Ltd., Publication, 2010).
[17] V. Tsakova, *J. Solid State Electrochem*., 12 (2008) 1421.
[18] N. S. Lawrence, H.-P. Liang, Metal Nanoparticles: Applications in Electroanalysis, *Nanostructured Materials in Electrochemistry*, Ed. A. Eftekhari. Weinheim: Wiley-VCH, (2008) 435.
[19] A. J. Killard, Nanostructured Conducting Polymers for Electrochemical Sensors, 565. (In book "Nanostructured Conducting Polymers" Ed. by A. Eftekhani, A John Wiley and Sons, Ltd., Publication, 2010).
[20] D. W. Hatchett and M. Josowicz, *Chem. Reviews*, 108 (2008)746.
[21] M. A. Vorotyntsev and S.V. Vasilyeva, *Adv. Colloid Interf. Sci*., 139 (2008) 97.
[22] R. Gangopadhyay and A. De, *Chem. Mater.,* 12 (2000) 608.
[23] Malinauskas, J. Malinauskiene, and A. Ramanavicius, *Nanotechnology,* 16 (2005) 51.
[24] B. C. Sih and M.O. Wolf, *Chem. Commun*., (2005)3375.
[25] V. V. Malev, V. V. Kondratiev, A. M. Timonov, Polymer-Modified Electrodes, Nestor-Istoriya, St. Petersburg, 2012, 365 (in Russian).
[26] S. Roth, One-Dimensional Metals, VCH, Weinheim (1995)

[27] R. R. Chance, D.S. Boudreaux, J. L. Bredas, R. Silbey, in: T. A. Skotheim (Ed.), *Handbook of Conducting Polymers*, vol. 2, Marcel Dekker, New York, (1986) 825.
[28] J. L. Bredas, G. B. Street Polaron, Bipolarons and Solitons in Conducting Polymers. *Acc. Chem. Res.* 18 (1985) 309.
[29] J. L. Bre´das, J.C. Scott, K. Yakushi, and G.B. Street, Polarons and bipolarons in polypyrrole:evolution of the band structure and optical spectrum upon doping, Phys. Rev. B.: *Condens. Matter*., 30(1984) 1023.
[30] K. Mizoguchi, S. Kuroda, in: H.S. Nalwa (Ed.), *Handbook of Organic Conducting Molecules and Polymers,* vol. 3, Wiley, New York, (1997) 251.
[31] M. Nechtschein, F. Devreux, F. Genoud, E. Vieil, J. M. Pernaut, E. *Genies, Synth. Met*., 15 (1986) 59.
[32] M. F. Mathias, O. Haas, *J. Phys. Chem*., 96 (1992) 3174.
[33] M. M. Musiani. *Elchim. Acta*, 35 (1990)1665.
[34] M. A. Vorotyntsev, E Vieil, J Heinze, J. *Electroanal. Chem*., 450 (1998) 121.
[35] M. A. Vorotyntsev,L I Daikhin, M D Levi *J. Electroanal. Chem*., 364 (1994) 37
[36] M. A. Vorotyntsev, J-P Badiali, G Inzelt *J. Electroanal. Chem*., 472 (1999) 7
[37] M. A. Vorotyntsev, C Deslouis, M M Musiani, B Tribollet, K Aoki *Electrochim. Acta* 44 (1999) 2105
[38] V. V. Malev, A. A. Rubashkin, M. A. Vorotyntsev *Rus. J. Elchem*., 33 (1997) 945.
[39] V. V. Malev, V. V. Kondratiev *Russian Chemical Reviews* 75 (2006)147.
[40] M A Vorotyntsev, *Electrochim. Acta*, 47(2002) 2071.
[41] A. F. Diaz, J. Bargon, in: T.A. Skotheim (Ed.), *Handbook of Conducting Polymers*, vol. 1, Marcel Dekker, New York, (1986) 81.
[42] G. Tourillon, in: T. A. Skotheim (Ed.), *Handbook of Conducting Polymers,* vol. 1, Marcel Dekker, NewYork, (1986) 293.
[43] J. Tanguy, J. L. Baudoin, F. Chao, M. *Costa Electrochim*. Acta, 37 (1992) 1417.
[44] J. Tanguy, A. Pron, M. Zagorska, I. Kulszewicz-Bajer. *Synth. Met*., 45 (1991) 81.
[45] K. Juttner, R. J. H. Schmitz, A. Hudson, *Electrochim. Acta*, 44 (1999)4177.
[46] S. A. M. Refaey *Synth. Met*., 140 (2004)87.

[47] S. A. M. Refaey, G. Schwitzgebel Design. *Monomers Polym.*, 3 (2000) 389.
[48] O. A. Semenikhin, E. V. Ovsyannikova, N. M. Alpatova, Z. A. Rotenberg J. *Electroanal. Chem.*, 408 (1996) 67.
[49] V. V. Kondratiev, E. G. Tolstopyatova, Ya. V. Trofimova, V. V. Malev, *Elektrokhimiya* 39 (2003) 751.
[50] M. Grzeszczuk, J. Bobacka, J. Aivaska, *Electroanal. Chem.*, 362 (1993) 287.
[51] J. Bobacka, M. AIvaska, *Grzeszczuk Synth. Met.*, 44 (1991)9.
[52] J. Bobacka, M. Grzeszczuk, Aivaska Electrochim. *Acta*, 37 (1992)1759.
[53] C. Kvarnstrom, AIvaska *Synth. Met., 84* (1997) 593.
[54] E. G. Tolstopyatova, S. N. Sazonova, V. V. Kondratiev, V. V. Malev, *Rus. J. Elchem.* 40 (2004)1073.
[55] J. Bobacka, A. Levenstam, A. Ivaska *J. Electroanal. Chem.*, 489 (2000) 17.
[56] H. Randriamahazaka, V. Noel, C. Chevrot, *J. Electroanal. Chem.*, 472 (1999) 103.
[57] H Randriamahazaka, V Noel, C Chevrat *J. Electroanal. Chem.,* 521 (2002) 107.
[58] H Randriamahazaka, V Noel, C Chevrat *J. Electroanal. Chem.*, 556 (2003)35.
[59] G. A. Snook, C. Peng, D. J. Fray, G. Z. Chen, *Electrochem. Commun.*, 9 (2007) 83.
[60] E. Laviron, Electroanalyt. *Chem. Interfacial Electrochem.*, 52 (1974) 355.
[61] E. Laviron, *Electroanalytical Chemistry*, V.12, P. 53, ed. A.J. Bard, Marcel Dekker, N.Y., 1982
[62] V. V. Malev, O.V Levin, *Electrochimica Acta.*,80 (2012) 426.
[63] V. V. Malev, O.V. Levin, V.V. Kondratiev. *Elchim. Acta.,* 122 (2014) 234.
[64] R. L. Hurt, J. R. Macdonald, *Solid State Ionics*, 20 (1986)111.
[65] M. Orazem, T E Moustafid, C Deslouis, B Tribollet *J. Electrochem. Soc.*, 143 (1996) 3880.
[66] C. Deslouis, T. E. Moustafid, M. MMusiani, B. Tribollet *Electrochim. Acta,* 41 (1996) 1343.
[67] P Ferloni, M Mastragostino, L Meneghello *Electrochim. Acta*, 41 (1996) 27.
[68] G. Garcia-Belmonte, J. Bisquert, E. C. Pereira, F. Fabregat- Santiago. *J. Electroanal. Chem.*, 508 (2001) 48.

[69] J. Bisquert, *Electrochim. Acta,* 47(2002) 2435.

[70] C. Ehrenbeck, K. Juttner, S. Ludwig, G. Paasch, *Electrochim. Acta*, 43(1998) 2781.

[71] H. Ding, Z. Pan, L. Pigani, R. Seeber, C. Zanardi, *Electrochim. Acta*, 46 (2001) 2721.

[72] H. Ding, Z. Pan, L. Pigani, R. Seeber *J. New Mater. Electrochem. Syst.* 4 (2001)61.

[73] B. W. Johnson, D. C. Read, P. Christensen, A. Hamnett, R. D. Armstrong, *J. Electroanal. Chem.*, 364 (1994)103.

[74] E. G. Tolstopyatova, S. N. Sazonova, V. V. Malev, V. V. Kondratiev, *Electrochim. Acta* 50 (2005) 1565.

[75] S. N. Eliseeva, D. V. Spiridonova, E. G. Tolstopyatova, V. V. Kondratiev, *Russ. J. Electrochemistry*, 44 (2008) 894.

[76] S. H. Glarum, J. H. Marshall, *J. Electrochem. Soc.*, 134 (1987) 142.

[77] Rubinstein, E. Sabatani, J. Rishpon, *J. Electrochem. Soc.*, 134(1987) 3078.

[78] T. Komura, T. Yamaguti, K. Takahasi *Electrochim. Acta*, 41 (1996) 2865.

[79] T. Komura, Y. Funahasi, T. Yamaguti, K. Takahasi, *J. Electoanal. Chem.* 446 (1998) 113.

[80] K. Martinusz, G. Lang, G. Inzelt, *J. Electroanal. Chem.* 433 (1997)1.

[81] R. I. Tucceri, C. Barbero, J. J. Silber, L. Sereno, D. Posadas, *Electrochim. Acta*, 42 (1997) 919.

[82] D. Goncalves, R. C. Faria, M. Yonashiro, L. O. S. Bulhoes, *J. Electroanal. Chem.*, 487 (2000) 90.

[83] K. Malek, F. Gobal, *Synth. Met.,* 113 (2000) 167.

[84] C. Barbero, R. I. Tucceri, D. Posadas, J. J. Silber, L. Sereno, *Electrochim. Acta,* 40 (1995) 1037.

[85] F. J. Rodriguez- Nieto, D. Posadas, R. I. Tucceri, J. *Electroanal. Chem.*, 434 (1997) 83.

[86] T. Komura, Y. Ito, T. Yamaguti, K. Takahasi, *Electrochim. Acta,* 43 (1998) 723.

[87] F. J. RodrõÂguez Nieto, R. I. Tucceri, J. *Electroanal. Chem.*, 416 (1996) 1.

[88] E. M. Genies, E. Vieil, *Synth. Met.*, 20 (1987)97.

[89] S. N. Hoier, S-M Park, *J. Phys. Chem.*, 96 (1992)5188.

[90] K. Meerholz, J Heinze, *Electrochim. Acta*, 41 (1996) 1839.

[91] M. A. Vorotyntsev, J P Badiali, *Electrochim. Acta*, 39 (1994) 289.

[92] S. W. Feldberg, *J. Am. Chem. Soc*. 106 (1984) 4671.

[93] Y. Tezuka, K. Aoki, K. Shinozaki, *Synth. Met.* 30 (1989)369.
[94] V. Z. Barsukov, S Chivikov *Electrochim. Acta*, 41 (1996) 1773.
[95] K. Meerholz, J. Heinze Angew. *Chem., Int. Ed. Engl.,*29 (1990) 692.
[96] A. Smie, J. Heinze, Angew. *Chem., Int. Ed. Engl.*, 36 (1997) 363.
[97] A.A. Nekrasov, V. F. Ivanov, A. V. Vannikov, *J. Electroanal. Chem.*, 482 (2000) 11.
[98] A. Nekrasov, V. F. Ivanov, A. V. Vannikov, *Electrochim. Acta*, 46 (2001) 4051.
[99] J. Bisquert, G. Garcia-Belmonte, F. Fabregat-Santiago, N. S. Ferriols, M. Yamashita, E. C. Pereira, *Electrochem. Commun.*, 2 (2000) 601.
[100] J. Bisquert, G. Garcia-Belmonte, F. Fabregat-Santiago, P. R. Bueno, J. *Electroanal. Chem.*, 475 (1999)152.
[101] W. J. Albery, C. M. Elliott, A. R. Mount, J. *Electroanal. Chem.*, 288 (1990) 15.
[102] W. J. Albery, A. R. Mount, J. Chem. Soc., *Faraday Trans.* 89 (1993)327.
[103] W. J. Albery, A. R. Mount, *J. Electroanal. Chem.*, 388 (1995) 1.

3. UV-VIS-SPECTROELECTROCHEMISTRY OF PEDOT-MODIFIED ELECTRODES

A priory assumed connection between the electrochemical properties of the modified electrodes and their structural or morphology features requires an implication of various structural-physical studies. In that connection, combinations of purely electrochemical methods and such techniques, as UV-Vis-NIR-spectroelectrochemistry or electrochemical quartz crystal microbalance (EQCM) seem to be perspective, since their application might provide an additional information on the mass and charge transport processes in CPs. Actually, at studying of complicated *p*-doping processes with formation of quasi-particles that possess specific optical and electrochemical properties, a parallel and simultaneous use of electrochemical and spectroscopic characterization of polymer films leads, as known, to the necessary information. Such spectroelectrochemical studies in UV-visible range of electronic absorption spectra have been performed for different CPs, confirming the polaron-bipolaron concept of their conductance and the presence of detectable quasi-particles in the film interior. Most simple and easily treated data have been obtained, for instance in case of thiophene polymer family. In the papers related to such research [1-8], it is reported that electronic absorption spectra of polythiophene films contain two isobestic points. The latter means the presence of two redox transitions and three kinds of optically different absorbing particles in the systems studied. Let us consider in more detail the concrete examples of such studies performed with poly-3,4-ethylenedioxythiophene films in a number of works (see, e.g., [1-10]).

Figure 3.1 shows a series of steady–state electronic absorption spectra taken of a solid-state PEDOT film in a 0.1 M $LiClO_4$ solution at different electrode potentials [8]. The arrows indicated in the figure designate the direction of the optical density variation at a gradual film oxidation, i.e. at increasing the positive potentials.

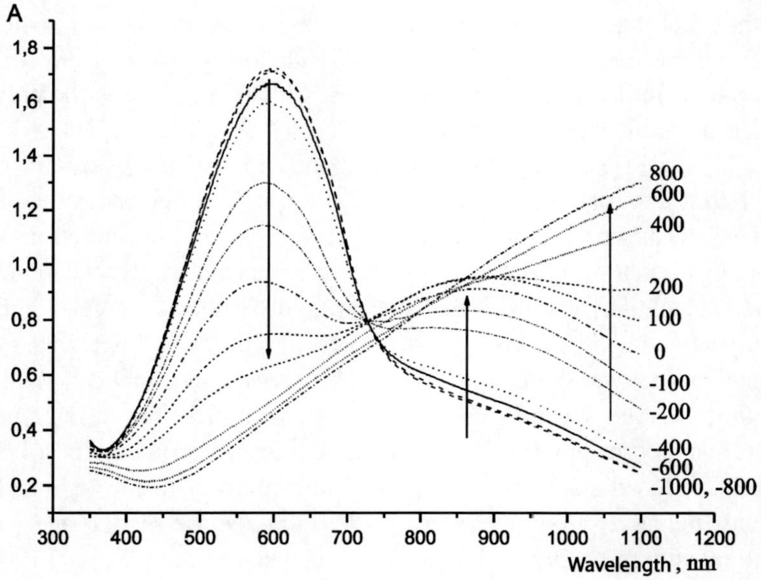

Figure 3.1. Electronic adsorption spectra of PEDOT film in 0.1 M $LiClO_4$ acetonitrile solution at different potentials, mV: -1000÷ 800.

As seen from Figure 3.1, the absorption bands of the reduced state (λ_{max} of about 600 nm, see below) coincide to each other at the most negative electrode potentials (–1000 and –800 mV), which indicates to the constant concentration of the reduced fragments in this potential range. In accordance with the literary data on PEDOT films [1, 2], this absorption band can be attributed to π-π* interband electronic transitions in neutral film fragments. Wavelength λ_{max} of the absorption maximum of the PEDOT reduced form equals 600 nm, which is in a good agreement with the literature data [3–6]. A gradual decrease in the intensity of the π-π* electronic transition band at λ_{max} = 600 nm and the observed shift of the absorption peak maximum towards the short–wave range occur in the course of the polymer oxidation (or doping), that is, a hypsochromic shift of the peak is registered at the conditions indicated. As can be seen from Figure 3.1, the absorption band intensity at λ_{max} = 600 nm practically stops its changing at the highest positive potentials.

An increase in the positive electrode potential results in the simultaneous appearance of the second and wide absorption band with a maximum at about λ_{max} = 850 nm. Its intensity grows at the potential increase up to E = 200 mV. At higher potentials, the absorption growth slows down and the edge of a new absorption band appears in the near IR spectral range with a maximum probably located beyond the registered wavelength range (λ_{max} > 1100 nm). It is worth to not that the above evolution of the spectra is typical for different thiophene derivatives at variations in their oxidation degree [1-7, 9, 10]. This indicates to a similar nature of the processes, which take place in the polymers with their different chemical substitutes.

The observed existence of isobestic points in the set of spectra of the polymer film at variations in the film oxidation degree evidences that, in the case of PEDOT films, we deal with the constant overall concentration of two absorbing polymer fragments. Their ratio appears to be dependent on the electrode potential. Besides, the presence of isobestic points indicates indirectly to a compliance with the Beer's law for absorbing fragments.

Thus, one can speak on at least three absorption bands that can be singled out in the observed spectrum. The first one with a pronounced maximum at λ_{max} = 600 nm corresponds to π-π* electronic transitions (in case of a PEDOT film in its reduced state), but other two ones conform to the oxidized film fragments: the absorption band at λ_{max} = 850 nm and the second one with its possible maximum beyond the registered wavelength range (λ_{max}> 1100 nm). It should be noted that the observed evolution of electronic absorption spectra of PEDOT films as a process dependent on the applied potential is a common phenomenon, which earlier observed for other polythiophenes [7, 11-15].

So the above data indicate to the appearance in the PEDOT film of three kinds of absorbing species and this is in agreement with the well-known polaron-bipolaron model of redox processes in CPs. These redox processes can be represented by two consecutive electrochemical stages, each of which possesses its certain formal potential (E_1^0, E_2^0):

$$R1 \rightleftarrows O1 + e \qquad E_1^0$$
$$O1 \rightleftarrows O2 + e \qquad E_2^0 \tag{3.1}$$

Here, symbol R1 obviously corresponds to the neutral fragments of PEDOT films, whereas symbols O1 and O2 should most probably be assigned to polaron and bipolaron species.

References to Section 3

[1] C. Carlberg, X. Chen, and Inganaes O., *Solid State Ionics*, 85 (1996) 73.
[2] X. Chenand O. Inganaes, *J. Phys. Chem.*, 100 (1996) 15202.
[3] C. Kvarnstrom, H. Neugebauer, S. Blomquist, H.J. Ahonen, J. Kankare, and A. Ivaska, *Electrochim. Acta*, 44 (1999) 2739.
[4] S. Garreau, G. Louarn, J.P. Buisson, G. Froyer, and S. Lefrant, *Macromolecules,* 32 (1999) 6807.
[5] M. Lapkowski and A. Pron,, *Synth. Met.*, 110 (2000) 79.
[6] T. Tung and K. Ho, *Solar Energy Materials and Solar Cells*, 90 (2006) 521.
[7] V. V. Kondratiev, V. V. Kurdakova, N. G. Antonov, E. G. Tolstopyatova, *Russ. J. Electrochem.,* 44 (2008) 286.
[8] E.G. Tolstopyatova, N.A. Pogulaichenko, S.N. Eliseeva, V.V. Kondratiev *Russ. J. of Electrochem.,* 45 (2009) 252.
[9] N. M. Alpatova, Z. A. Rotenberg, E. V. Ovsyannikova, V.V. Topolev, M. Yu. Grosheva, S. Kirchmeyer, and F. Jonas, *Russ. J. Elektrochem.*,40 (2004) 917.
[10] Alpatova, N. M., Ovsyannikova, E. V., Topolev, V. V., and Grosheva, M. Yu., *Russian J. of Electrochemistry*, 40 (2004) 229.
[11] Li Niu, C. Kvarnstrom, K. Froberg, and A. Ivaska, *Synth. Met.*, 122 (2001) 425.
[12] S. N. Hoier, and S.-M. Park, *J. Phys. Chem.*, 96 (1992) 5188.
[13] M. Trznadel, M. Zagorska, M. Lapkowski, G. Louarn, S. Lefrant, and A. Pron, *J. Chem. Soc.*, Faraday Trans., 92 (1996) 1387.
[14] R.-M. Latonen, C. Kvarnstrom, and A. Ivaska, *J. Electroanal. Chem.*, 512 (2001) 36.
[15] M. Skompska, *Electrochim. Acta*, 44 (1998) 357.

4. EQCM Investigations of Mass Transfer Processes in PEDOT Films

The electrochemical quartz crystal microgravimetry (EQCM) method has played an important role in studying the electrode processes in CPs. In many cases from such measurements, one can extract a reliable knowledge on the nature of counter-ions incorporated into a polymer film and changes in its viscoelastic properties during the course of an electrode process. The *p*-doping processes occurring in the CP films include formation of positive charges in polymer chains that is accompanied with counter-ions and solvent molecules transfer into the polymer bulk and, probably, some conformation changes. The regularities of these complex processes determine electrochemical and microbalance behavior of these systems.

In case of acoustically thin films, EQCM measurements of shifts of the resonance frequency of the electrode crystal oscillations allow one to obtain conclusions as to the average molar mass of transferred particles participating in the electrode process. In particular, microgravimetrical responses of PPy-modified electrodes and some poly-3-alkylthiophenes ones demonstrate typical monotonous increases of the electrode mass with increasing the film oxidation level (Figure 4.1).

These dependencies are in agreement with uptake of anions–dopants during the oxidation of films and release of the same ions at the reverse process. The calculated values of the apparent molecular mass of transferred species were close to those of anions of the electrolytes used. Some ingress or a lack of the transferred mass as compared to the molecular mass of the electrolyte anion is usually assigned to the arising fluxes of solvent molecules during the doping process. However in many cases, a more complicated behavior is observed, which results from non-monotonous changes in the resonance frequency with charging the films and a possible participation of both cations and anions in the studied electrode processes.

In particular, more complicated microgravimetrical responses of PEDOT films were observed in [1-7]. Though these films are often considered as reference polymer ones due to stability of their electrochemical properties in a wide range of potentials, anyone can note a great difference in the behavior of PEDOT films and other alkyl-substituted polythiophenes. Such difference in the film microgravimetric responses is reflected with the appearance of different regions of the mass dependence on the electrode potential, the effect probably caused by conformation changes occurring in the films at potential variations. The observed dependence of resonance frequency *f* on the electrode potential (in the range of – 0.8 to 0.8 V) reveals two its different regions at using solutions of $TBABF_4$ and $TBAPF_6$ in PC (corresponding curves are not given here).

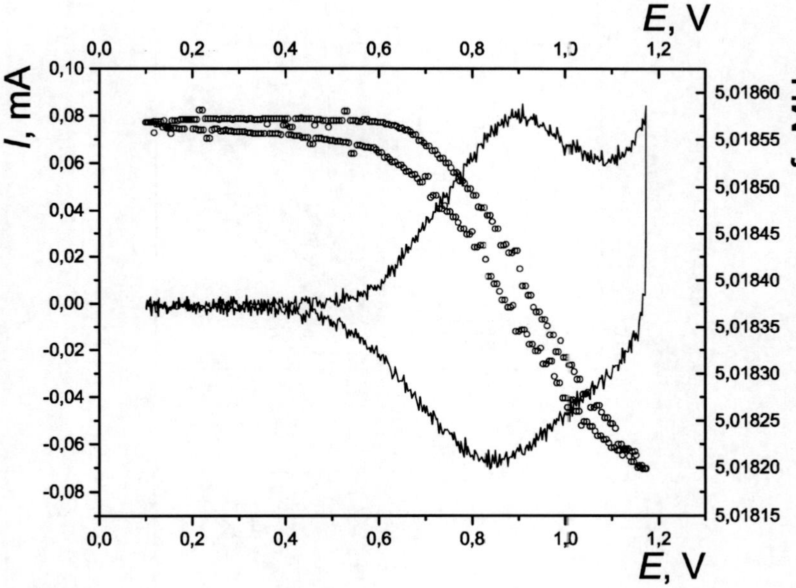

Figure 4.1. CV of POT film (line) and corresponding frequency-potential dependence (dots) for the POT modified quartz crystal electrode. 0.1M $LiClO_4$ acetonitrile solution.

The first and most complex region (I) is observed in the potential range of – 0.8 to 0.3 V. In this region, the resonance frequency increases at a gradual rise of the potential and this conforms to a film mass decrease. A similar region of the *f*(*E*)-dependence was pointed out in [3] for acetonitrile solutions of $TBAPF_6$, though an increase of the resonance frequency was much less pronounced at the initial stage of the PEDOT film oxidation. The second characteristic region (II) of the dependence is in the range of more positive potentials of 0.3 – 0.8 V. A monotonous frequency decrease in region (II) at rising the electrode potential points to the film mass increase resulted from entering the counter-ions into the film interior. The direct and reverse runs of the *f(E)*-dependence do not coincide, i.e. some hysteresis phenomenon is observed.

Effects of the electrolyte ions nature were also investigated to elucidate stoichiometric mechanisms of redox processes in PEDOT films. The obtained Δm vs. ΔQ curves of PEDOT films are represented in Figure 4.2a for perchlorate solutions containing Li^+, Na^+, and TBA^+-ions. The anion effects on the Δm, ΔQ-dependence are illustrated by Figure 4.2b.

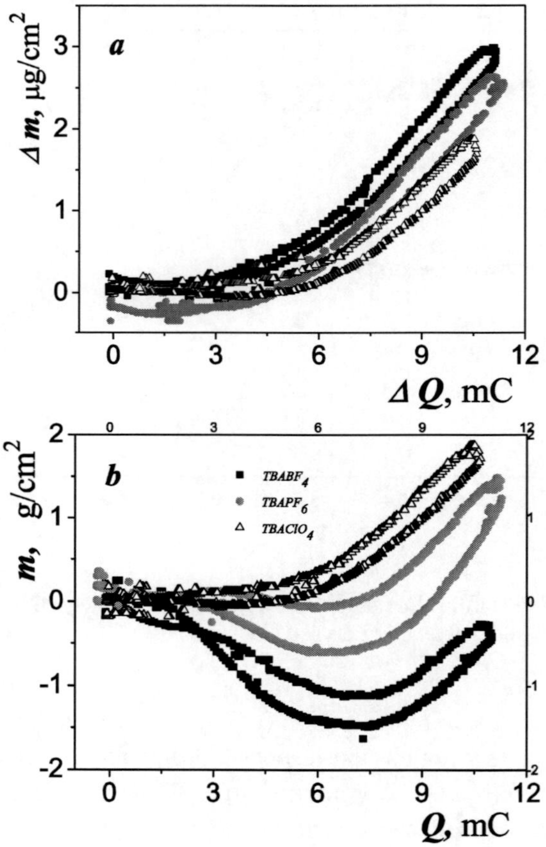

Figure 4.2a,b. Mass-qoulombic curves for PEDOT films in 0.1 M salts PC electrolytes: a – (•) $LiClO_4$, (•) $NaClO_4$ and (Δ) $TBAClO_4$; b – (•) $TBABF_4$, (•) $TBAPF_6$ and (Δ) $TBAClO_4$.

As it is seen from the figures, significant effects of the nature of the supporting electrolyte ions of both signs are observed in the mass change curves. The film mass somewhat decreases passing through a minimum or remains practically constant in the initial region, though the electricity consumed is rather significant (up to 30% of the overall electricity amount). In the second region of the $m(Q)$-dependence (corresponding to the potential range of 0 – 0.8 V), a monotonous (and practically linear) mass increase is observed at an increase in the electricity amount consumed during the anodic cycling. The same shape of the curves is also registered in this region at the reverse cathodic cycle of the film reduction.

The anion effect on the Δm, ΔQ-dependence in the case of a PEDOT film is illustrated by Figure 4.2b. At the first step of oxidation (corresponds to the potential range of –0.8 to 0.1 V) in PC solutions of $TBABF_4$ and $TBAPF_6$, the film mass decreases and then it becomes to increase.

The observed character of mass variations within the potential range of 0–0.8 V qualitatively agrees with that corresponds to anion-doping processes. The average molar mass values for transferred particles are found by using the slopes of $m(Q)$-dependences according to the Sauerbrey equation:

$$\Delta f = - C_f \cdot \Delta m, \tag{4.1}$$

where Δf, the observed shift of the resonant frequency (Hz); Δm, the corresponding mass change (g/cm^2); C_f, the sensitivity factor of the used crystal.

As indicated in Table 1, these values for solutions of lithium, sodium, and tetrabutylammonium perchlorates occur be equal to 62, 61, and 57 g/mol, correspondingly, each of which is considerably lower than the molar mass of non-solvated perchlorate-ions (99.5 g/mol).

Table 1. Experimental values of the average molar mass of the mobile species per 1 F of the consumed electricity in PEDOT films for solutions of different electrolytes. The data were calculated from mass change curves registered at the potential sweep rate of 0.01 V/s

Electrolyte	M_{A^-}(calculated), g/mol	M(exper.), g/mol	N solvent molecules
$LiClO_4$	99	62 ± 3	0.3–0.4
$NaClO_4$	99	61 ± 3	0.3–0.4
$TBAClO_4$	99	57 ± 3	0.5–0.6
$TBABF_4$	87	45 ± 2	0.3–0.4
$TBAPF_6$	145	65± 4	0.7-0.8

The decreased molar mass values as compared to the anion mass point to the simultaneous removal of the solvent molecules from the film in the course of the film oxidation. As seen from the table, the number of solvent molecules leaving the film varies from 0.3 up to 0.8 per a single anion entering the film [7]. Thus in all the cases, an underestimated average molar mass of transferred anions is observed. This evidences the opposite directions of the anion and solvent fluxes. Such conclusion agrees with the data obtained in [1] for PEDOT films in an acetonitrile solution of lithium perchlorate.

The authors of [1] showed that variations in the degree of the film filling with the solvent have not changed monotonously in the course of *p*-doping; firstly, the film oxidation results in a significant decrease of the solvent amount, then the solvent flux changes its direction, when it enters the film, and after a film desolvation again occurs at more high positive potentials. Thus, the reduced film form seems to be the most solvated one and the solvent and ion-dopant fluxes are not interrelated.

The same conclusions concerning to the solvent transfer from poly-3-alkylthiophene films in the presence of hexafluorophosphate anions have been done in papers [8, 9]. The authors of works [9, 11] assume that two types of polymer zones exist in polythiophene films: the zones with a high conjugation length and the regions with its lower value. The zones of the first type are more compact than those of the second one and practically desolvated, while the second zones have a more open configuration and easily include the solvent molecules. The latter can be replaced by anions in the oxidation process. In the reverse cathodic cycle, the solvent species enter the film in the considered potential region [1, 2]. In paper [2], the mass transfer processes in PEDOT films are also discussed for both the aqueous and acetonitrile solutions containing a number of different salts. Its authors also conclude that the anion and solvent fluxes are only responsible for the observed mass changes.

A different conclusion was made in [3], where analysis of EQCM data for acetonitrile solutions of a number of electrolytes has been performed basing on a model that accounts for the possibility of a common participation of both cations and anions in *p*- and *n*-doping processes. The authors of [3] believe that the electrolyte cations (in particular, large tetrabutylammonium ones (TBA^+)) participate together with anions in the *p*-doping process of PEDOT films. The cation transport into the films is predominant at the potentials below – 0.2 V, while the anion one begins at higher potentials. This conclusion is also supported by the data obtained in work [4], where ion transfer has been studied by EQCM in aqueous solutions containing such cations, as Na^+, K^+, and Ru^{+3}. The mixed participation of the electrolyte ions in charge transfer processes within PEDOT films is also discussed in papers [5, 6].

As it took place under the preceding discussion of the electrochemical results obtained with CPs, the above data have also created some impression on complexity of the phenomena related to charge transfer within these objects. Probably, the most important result following from microgravimetrical studies of CPs is in the appearance of solvent fluxes that accompany charge transport within the films under consideration. This fact directly shows the existence of interactions between polymer chains and surrounding solvent molecules. Moreover, it indicates to some changes in such interactions at charging/discharge of the studied films. It remains only to add here that the presence of these interactions and the possibility of their changes are not accounted for by the existing theory of modified electrodes.

References to Section 4

[1] A. R. Hillman, S. J. Daisley, S. Bruckenstein, *Electrochem. Com.*, 9 (2007) 1316.

[2] Bund, S. Neudeck, *J. Phys. Chem. B*, 108 (2004) 17845.

[3] Li Niu, C., Kvarnstrom, A. Ivaska, *J. Electroanal. Chem.*, 569 (2004) 151.

[4] N. Yang, C. Zoski, *Langmuir,* 22 (2006) 10338.

[5] F. Blanchard, B. Carre, F. Bonhomme, P. H. Biensan, D. J. Lemordant, *Electroanal. Chem.,* 569 (2004) 203.

[6] H. Randriamahazaka, C. Plesse, D. Teyssie, C. Chevrot, *Electrochem. Com.,* 6 (2004) 299.

[7] S. N. Eliseeva, T. A. Babkova, V. V. Kondratiev *Russian J. of Electrochemistry*, 45 (2009) 152.

[8] V.V. Kurdakova, N. G. Antonov, V. V. Malev, V. V. Kondratiev, *Russ. J. Electrochem.,* 42 (2006) 299.

[9] C. Visy, J. Kankare, E. Krivan, *Electrochim. Acta*, 45 (2000) 3851.

[10] M. Skompska, A. Jackson, A. R. Hillman, *Phys. Chem. Chem. Phys.*, 2 (2000) 4748.

[11] C. Visy, J. Kankare, *Electrochim. Acta*, 45 (2000) 1811.

5. Metal-Containing Composites Based on Conducting Polymers

5.1. Synthesis of Metal-Containing Polymers

As mentioned in the Introduction, a lot of research of nanocomposite materials formed from CPs has been performed during the last decade [1–9]. The strategy used at incorporation of metallic particles or other compounds into CP films is in the following. It is assumed that syntheses of these composites should lead to some positive effects related to the three-dimensional distribution of the included particles, the improved transport characteristics for electronic and ionic charges in the film matrix, and the increased catalytic activity of the inclusions due to producing of their nano-sized structures.

A number of different synthetic methods reviewed partly in refs. [1, 2] were proposed for preparation of such composites. These methods include either chemical or electrochemical deposition of metals from the corresponding solutions, incorporation of preliminary synthesized metal nanoparticles in the course of film electropolymerization, spontaneous reduction of metal ions by monomer or polymer film precursors, codepostion during potentiodynamic or pulse potential sweep, etc. Probably, the simplest way to load metal particles into CP films is electrodeposition of metals, which has been employed firstly for PMT [10] and PPy [11]. Aiming a fabrication of electrocatalytic materials, various metals were electrodeposited in CPs, in particular Au [12-18], Pd [19-30], Pt [31-70], Ag [71-75], etc. The metal electrodeposition can obviously be performed into a preliminary synthesized CP film or by simultaneous formation of a polymer film with included metal particles during one stage syntheses. In the last case, a common deposition of polymers and metal particles might proceed with using cyclic voltammetry in a wide range of electrode potentials in the presence of monomer film precursors and metal ions in a working solution. Asymmetric potential impulses can also be applied for the monomer oxidation and metal ions reduction at different chosen potentials. In this case, the oxidative polymerization and metal clusters growth take place at using the same working solution, but during the time-resolved processes.

It is clear that, depending on the technique applied for the metal electrodeposition, one can obtain significant differences in the size and spatial distribution of metal particles within the film volume. These issues are reviewed in detail [2], and the important role of the polymer morphology at such applications is clearly demonstrated in [29]. A possible polymer surface inhomogeneity and some energetic and topographic peculiarities of different CPs are usually considered as factors determining different sizes of the formed aggregates or their agglomeration [76-78]. Depending on the CP porosity and the film conductivity at the potentials of a metal deposition, different spatial distributions of metal particles across the film might take place [5]. In general, it is expected that, for poorly conducting and porous films, a metal deposition starts from an inner (substrate/CP) interface growing through the polymer layer. In the case of highly conducting polymers, nucleation of metals takes place predominantly at an outer (polymer/solution) interface including the internal surface of film pores filled with the electrolyte used [78,79]. Pre-soaking of CP layers with electrolytes containing metal ions and a use of the impulse deposition technique lead to more homogenous distributions of metal clusters within polymer films [5, 33, 50, 69].

The electrochemical polymerization of modifying films in the presence of pre-synthesized metal nano-particles (MeNPs) is also the widely accepted method to prepare the composite electrodes [31-34, 80-83]. Its advantage is in the presumed use of homogeneous distributions of MeNPs over their size.

A special attention has been paid to the synthetic approaches, according to which either a CP film on the whole or its monomer precursors existing in a working solution act as reducing agents with simultaneous formation of metal particles in the polymer film or polymer structures in the solution, respectively. Simultaneous redox reactions between ions of electropositive metals (mostly complex ions of noble metals) and reduced forms of polymer films were used to prepare CP composites including such metals, as gold [84-93], palladium [28-30, 94-108], silver [109-119]. This method was also applied to produce metal-containing PEDOT films in our recent papers [92, 93, 106-108, 119].

PEDOT/Pd electrodes were obtained by the palladium ions reduction from a working solution containing $PdCl_2$ of different concentrations in contact with the reduced PEDOT film. As the standard electrode potential of the Pd(II)/Pd(0) couple is about +0.85 V (S.H.E.) in a 0.5 M H_2SO_4 solution, palladium ions can act as effective oxidizing agents for the reduced form of PEDOT films. The corresponding reaction is schematically represented by the equation

$$Pd^{2+} + 2PEDOT^{0} \xrightarrow{+2A^{-}} Pd^{0} \cdot (PEDOT^{+\cdot})_2 \cdot 2A^{-}, \quad (5.1)$$

where $PEDOT^0$ is the reduced fragment consisting of three-four neighboring repeat units of the polymer chains. Different anion-dopants A^- existing in a working solution (both sulfate- and chloride-ions) can participate in this reaction.

The direct evidence of loading palladium particles into preliminary reduced PEDOT films was obtained at the subsequent use of energy-dispersive X-ray analysis. It should also be mentioned that, in CV curves of such composite electrodes immersed into a 0,1 M H_2SO_4 solution, characteristic peaks of the hydrogen sorption–desorption, as well as the peaks associated with formation and reduction of PdO have been observed, which are the additional evidences of the presence of Pd particles in the films prepared.

On the other hand, the weight changes accompanying the palladium deposition into the prepared composite electrodes were studied by quartz crystal microbalance method. These measurements were carried out with PEDOT films of 0.1–0.2 μm thickness, which made it possible to neglect any contribution from their viscoelastic properties to the observed resonance frequency (f) shifts. Figure 4.3a shows how the shape of $f(E)$-dependence has changed in result of successive palladium loadings (30 s each time) of the same film after it was reduced and then maintained in a 0.1 M H_2SO_4 solution with the palladium chloride concentration of $5 \cdot 10^{-4}$ M. As a result, the $f(E)$ curves undergo a parallel displacement to the region of lower frequencies, which directly indicates to an increase in the amount of the palladium incorporated. From the observed shifts of the resonance frequency, the corresponding changes in the film mass were found by using Sauerbrey equation (3) (see Figure 4.3b).

At the same time, if one assumes that the palladium deposition into the films proceeds in accordance with Eqn. (5.1), the amount of the deposited palladium can be estimated independently, basing on the Faraday law. To perform such estimations, one needs two sets of experimental data. First, CV curves of PEDOT-modified electrodes in solutions that do not contain palladium ions must be established. Second, the dependence of changing the electrode potential in the course of the metal loading must be registered from its initiation to its end, when the loading is interrupted. Integrating a CV current curve divided by its sweep rate with respect to the electrode potential from its initial value up to the final one, obviously, gives the amount of electricity consumed during the palladium deposition. This makes it possible to calculate the loaded mass of Pd according to Faraday's law (see the fourth column of Table 2).

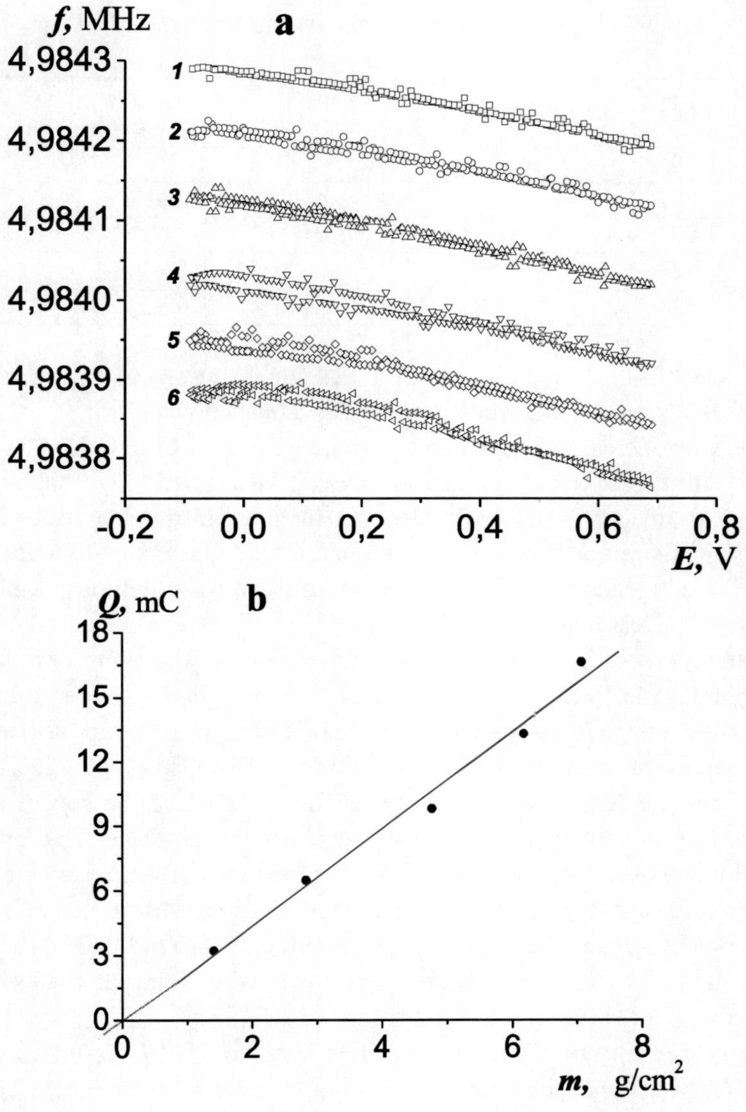

Figure 4.3. a) The resonant frequency of PEDOT (1) and PEDOT/Pd film (2-6) modified quartz crystal electrode as a function of potential (E) in 0.1M H_2SO_4. Curves 2-6 corresponds to successive 30 s deposition of Pd from $5 \cdot 10^{-4}$ M $PdCl_2$; b) Mass of palladium vs. amount of electricity passed during the oxidation of film by palladium ions.

Table 2. Loaded mass of Pd according to Faraday's law

№	Deposition time, τ (s)	$m_{Pd,}$ Δm (μg·cm^{-2}) EQCM	$m_{Pd,}$ Δm (μg·cm^{-2}) Coulombic method
1	30	1.41	1.30
2	60	2.83	2.61
3	90	4.77	4.46
4	120	6.18	5.90
5	180	7.17	6.82

As seen from the table, a good agreement between the results of both methods really takes place, though the mass values obtained from EQCM data are somewhat increased as compared to those calculated from CVs.

One of the reasons of such increase is the possibility that solution components (mainly water molecules) are included in the composite film in result of their sorption by palladium clusters. As the mass values obtained with two indicated means are sufficiently close, one can consider both methods be equivalent to each other.

Some morphologic features of PEDOT/Pd composite films were studied with scanning electron microscopy (SEM) [106,107]. Figure 5.1 represents the typical SEM image of their surface. One can see an irregular porous structure of the polymer film with a wide range of pore sizes. This figure also shows that the globular structures settled at the film surface and the pore walls are covered with mostly spherical nano-sized palladium particles. The latter are close in their size (10–20 nm for the chosen conditions of their deposition) and uniformly distributed over the polymer surface. More unambiguous evaluations of sizes of the deposited palladium particles and their distribution across the film thickness were performed with transparency electron microscopy (TEM) (Figures 5.2 and 5.3).

Figure 5.2 presents a bright-field TEM image of a PEDOT/Pd composite film at varying deposition time (deposition from $5 \cdot 10^{-3}$ M $PdCl_2$). In this recording mode, palladium particles are visualized as dark spots. As seen from Figure 5.2, the size of the dominant palladium clusters grows substantially with increasing the deposition time (as compared to their size at the deposition time equals 60 s). The data of Figure 5.2 confirm the volume character of palladium particles distribution (at least, in subsurface regions of the polymer); one can see a series of dense polymer regions surrounded by palladium particles.

Figure 5.1. SEM image of a PEDOT/Pd film. Pd particles are deposited by dipping of previously reduced polymer film for 60 s into $5\cdot10^{-3}$ M $PdCl_2$ / 0.1 M H_2SO_4 solution. Separate palladium clusters are indicated by arrows.

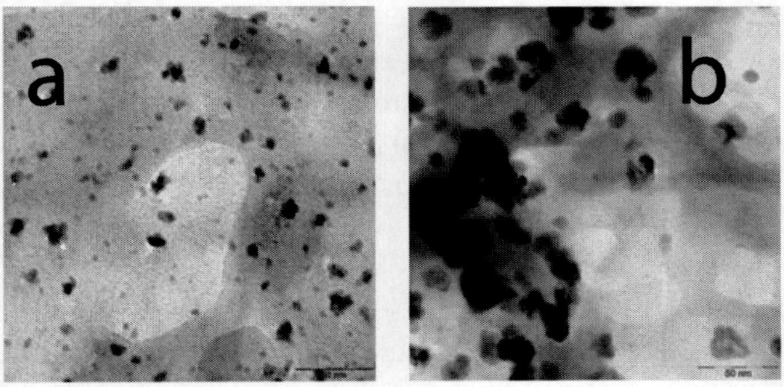

Figure 5.2. TEM images of a PEDOT/Pd film. a- 30 s deposition time ; b. 120 s deposition time from $5\cdot10^{-3}$ M $PdCl_2$ / 0.1 M H_2SO_4 solution.

Gold ions possessing of the high oxidation potential (1.002 V) can also be deposited into PEDOT films (and some other CP ones), producing metal particles in the film in result of the polymer fragments oxidation. This process is similar to the considered above, but schematically described by the equation:

$$3PEDOT^0 + AuCl_4^- + 3A^- = (PEDOT^+\cdot A^-)_3\cdot Au^0 + 4Cl^- \qquad (5.2)$$

Loadings of the metal gold particles account for their chemical deposition into PEDOT films have also been estimated by using the EQCM method and the approach based on the Faraday law. As it took place for PEDOT/Pd films, the same good correlation between the mass values obtained by two indicated means was established in case of PEDOT/Au films [92, 93]. The presence of such correlation permitted us to perform subsequent studies of redox processes on the electrodes modified with the above composite films.

5.2. Redox Reactions at Metal-Conducting Polymer Coated Electrodes

Metal clusters included into conducting polymers might accelerate numerous reactions such as oxidation of hydrogen, hydrogen peroxide, hydrazine (and other small organic molecules), reduction of oxygen, hydrogen peroxide, carbon dioxide, electrohydrogenation of organic species, and many others. Catalytic activity of metal nanoparticles in CPs has attracted much attention due to possible applications of these objects in low temperature fuel cells, catalysis of organic reactions, and sensors. An attractive feature of the metal–polymer composite materials is that nitrogen, oxygen, or sulfur atoms of polymer chains can stabilize nano-sized metal clusters due to interactions with the surface metal atoms, preventing agglomeration of such clusters into large-scale aggregates. Besides, such composite materials exhibit good transport properties due to their porosity and high electron conductivity.

In connection with such applications, the possible location of electrode reactions proceeding on the composites in question deserves a special discussion. In general, an electrode process with participation of redox-active particles may take place at the polymer/electrolyte interface, the phase boundary of metal clusters, and the inner electrode substrate/electrolyte interface in case of existing thorough film pores.

The experimental evidences of inhomogeneity of composite films resulted from their porosity and non-regular metal inclusions are shown in Figures 5.1-5.3. Therefore electrochemical reactions can actually occur not only on their visible "outer" surface, but also an "inner" surface of film pores. The porosity of composite films and the existence of different reaction surfaces (metal, substrate, and polymer) might lead to a complex electrochemical behavior of composite electrodes, which is not described with classical equations of the electrochemical kinetics on compact metal electrodes.

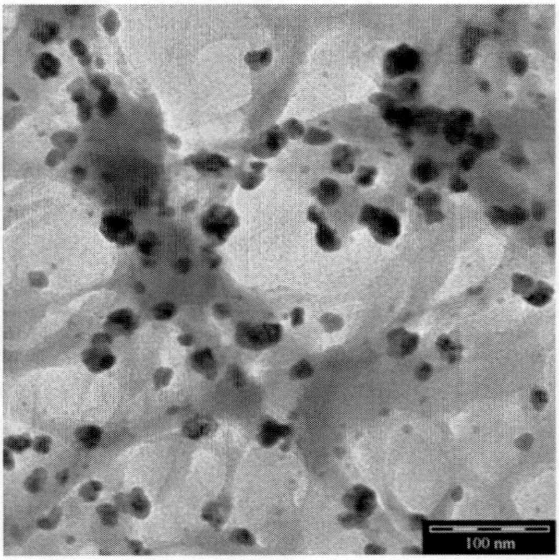

Figure 5.3. TEM image of PEDOT/Pd film (120 s palladium deposition time from $5 \cdot 10^{-3}$ M$PdCl_2$).

However, the performed quantitative simulations of electrode processes taking place in metal-containing films of CPs [121] show that a use of such equations often occurs possible if two essential conditions are fulfilled. The first one assumes that the heterogeneous rate constant of a studied process does not depend on the sizes of the metal clusters included into a CP film. The second condition is in a relative smallness of the film thickness as compared to the diffusion layer that appears in the adjacent solution under the passing currents. At the above conditions being fulfilled, the classical electrochemical equations (inclusive of non-stationary ones) appear to be valid for electrode processes in the metal-containing composite electrodes. The only point to be accounted for at such applications is the fact that the effective rate constant of the studied process occurs dependent on the metal loading into the used composite film (for more details see recent review [122]).

Two examples of experimental verifications of these theoretical inferences are shortly discussed below. In particular, the hydrogen peroxide reduction at PEDOT/Au composite electrodes [92] and the hydrazine oxidation at PEDOT/Pd composite film [107] have been studied as some testing reactions. It has been shown that, in 0.2 M phosphate buffer solutions (PBS) (pH = 6.86), the electroreduction of H_2O_2 proceeds simultaneously on both the gold clusters surface and the film one free from the metal inclusions. The observed shift of the half-wave potential of the H_2O_2 reduction on PEDOT/Au film (≈ - (500

÷ 570) mV) with respect to that of pristine PEDOT film (≈ -750 mV) amounts about 200 mV at the same experimental conditions. This indicates to the significant increase in the rate of the H_2O_2 electroreduction in the presence of dispersed gold particles incorporated into PEDOT films and, probably, results from electrocatalytic properties of the included gold particles with respect to the process under discussion.

Partial voltammetric curves of the hydrogen peroxide reduction on Au particles have been obtained and analyzed as functions of the electrode potential and the metal loading into PEDOT films [92]. As has been established, an increase in the amount of gold loaded into the films leads to a shift of the half-wave potential of the electroreduction wave to more positive potential values. At the same time, the observed limiting currents (j_d) increase their value and tend asymptotically to the limiting one that exactly corresponds to the Levich equation for diffusion limiting currents to rotating disc electrodes (RDE method has been applied in the study reviewed (see [92])). This phenomenon evidently results from a gradual transition from a small active surface of gold particles acting as separate microelectrodes at their low content in PEDOT films to the conditions of mass-transfer of reacting particles within the diffusion layer of the bathing electrolyte, when the amount of gold in composite films is high enough.

Based on the obtained partial voltammetric curves, the Tafel plots of E vs. log $[j/(j_d - j)]$ corrected for a mass transfer were constructed for the hydrogen peroxide reduction on the gold particles surface of composite films. Linear parts of such semi-logarithmic dependences were observed in the main range of the electrode potential changes (Figure 5.4), where the calculated partial currents were not too close to the observed limiting values. Tafel's slopes b_k of all the constructed curves were in the range of 0.21÷ 0.24 V/dec, which allowed us to conclude that an irreversible two-electron process took place in case of the hydrogen peroxide electroreduction on gold particles of PEDOT/Au films.

From the obtained slopes ($b_k = 2.303RT/\alpha' F$), the apparent transfer coefficient of the cathodic process in question was estimated as quantity $\alpha' = 0.24 \div 0.28$. The obtained value of α' indicates that the first electron transfer is a limiting step of the overall two-electron process. This is consistent with the two-steps hydrogen peroxide electroreduction suggested in [120]. In the case of a polycrystalline gold electrode, slope b_k of the linear part of the $E[\log(j/(j_d - j))]$-dependence occurs equal to 0.23 V/dec [92], i.e. it is close to that obtained for the hydrogen peroxide electroreduction on the surface of gold clusters included into PEDOT films.

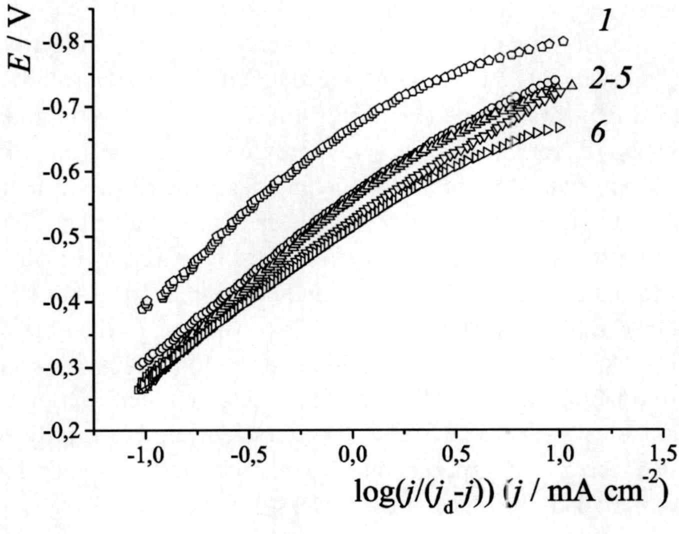

Figure 5.4. Mass-transfer corrected Tafel plots for the hydrogen peroxide reduction on Au-particles in PBS solution plus $1{\cdot}10^{-3}$ M H_2O_2 at different durations of gold loading, s: 1 - 60, 2 - 150, 3 - 240, 4 - 330, 5 - 560, 6 – 720.

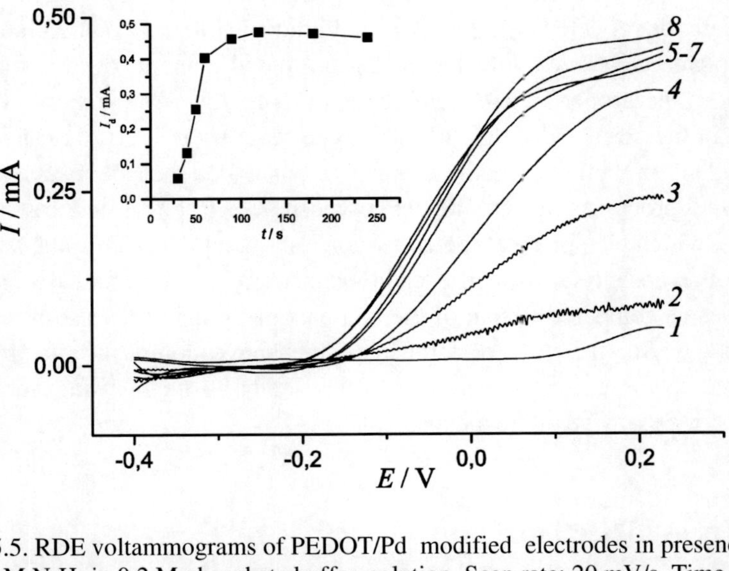

Figure 5.5. RDE voltammograms of PEDOT/Pd modified electrodes in presence of $2.5{\cdot}10^{-3}$ M N_2H_4 in 0.2 M phosphate buffer solution. Scan rate: 20 mV/s. Time of Pd deposition from $5{\cdot}10^{-3}$ M $PdCl_2$/0.1 M H_2SO_4, s: 1 – 30, 2 – 40, 3 – 50, 4 – 60, 5 – 90, 6 – 120, 7 – 180, 8 – 240. On the inset – dependence of limiting current I_d from deposition time.

Analogous approach has been applied in work [107] devoted to the hydrazine electro-oxidation on PEDOT/Pd-modified electrodes. An asymptotic growth of the limiting currents of hydrazine oxidation with increasing Pd deposition time (Figure 5.5) has been observed similar to that mentioned above for the hydrogen peroxide reduction on PEDOT/Au electrodes. As it took place in that preceding case, the diffusion nature of the limiting currents of the hydrazine oxidation on composite PEDOT/Pd films in PBS has been established. This allowed us to estimate the hydrazine diffusion coefficient in the adjacent solution as value equal to $1.2 \cdot 10^{-5}$ $cm^2 s^{-1}$, which is in agreement with the data published in [124]. Analysis of the obtained cyclic voltammetry data led us to the conclusion that the four-electron process of the hydrazine oxidation on PEDOT/Pd composites proceeded predominantly on the palladium clusters distributed over the surface layers of the polymer films.

Thus, the performed experimental studies with metal-containing PEDOT films show that their results might successfully be treated in scope of the traditional approaches to electrochemical processes on metal electrodes. As mentioned above, the only point to be accounted for at using such equations in the case of metal-containing CP films is in the appearance of some dependence of the effective rate constant of the studied process on the metal loaded into the used film. The existence of such dependence for the both systems investigated is reflected with the observed gradual shifts of polarization curves of composite electrodes with increasing the metal content towards the curve inherent to the proper metal electrode (see Figure 5.4). As the slopes of these curves (in the corrected Tafel coordinates: E vs.lg $[j/(1-j/j_d)]$) are practically independent on the metal loading, it can be concluded that the mechanisms of the studied processes in cases of the composite electrodes tested are the same as those, which take place at the corresponding metal ones. This allows one to assume the absence of specific catalytic activity of the studied composite electrodes as compared to that of the compact metal ones, at least in the cases investigated. So, the only preference of the above composite electrodes as compared to the corresponding solid ones seems to be in using very small amounts of expensive noble metals.

References to Section 5

[1] B. C. Sih, M. O. Wolf, *Chem. Commun.*, (2005), 3375.

[2] V. Tsakova, Metallic-Based Nanocomposites of Conductive Polymers P.289. In book "*Nanostructured Conducting Polymers*" Ed. by A. Eftekhani, A John Wiley and Sons, Ltd., Publication, 2010
[3] X. Li, J. Sun, and M. Huang, Preparation and properties of nanocomposites of polyaniline and metal nanoparticles, Progr. Chemistry, 19 (2007)787.
[4] D. W. Hatchett and M. Josowicz, Composites of intrinsically conducting polymers as sensing nanomaterials, *Chem. Reviews,* 108 (2008) 746.
[5] V. Tsakova, How to affect number, size and location of metal particles deposited in conducting polymer layer, *J. Solid State Electrochem*., 12 (2008)1421.
[6] R. Gangopadhyay and A. De, Conducting polymer nanocomposites: A brief overview. *Chem. Mater*., 12 (2000) 608.
[7] Malinauskas, J. Malinauskiene, and A. Ramanavicius, Conducting polymer-based nanostructurized materials: electrochemical aspects, *Nanotechnology*, 16 (2005) R51
[8] E. Antolini and E.R. Gonzalez, Polymer supports for low-temperature fuel cell catalysts, *Appl. Catal.* A, 365 (2009) 1.
[9] C. Lamy and J.M. Leger, Electrocatalysis with electron conducting polymers modified by noble metal nanoparticles. In "*Catalysis and Electrocatalysis at Nanoparticle Surfaces*", ed. Vieckowski, E. Saviniva, and C. Vayenas, Marcell Dekker, New York, 2003.
[10] G. Tourillon and F. Garnier, Inclusion of metallic aggregates in organic conducting polymers, *J. Phys. Chem*., 88 (1984) 5281.
[11] G.K. Chandler and D. Pletcher, *J. Appl. Electrochem*., 16 (1986) 62.
[12] J.M. Kinyanjui, J. Hanks, D.W. Hatchett, A.J. Smith, and M. Josowicz, *J. Electrochem. Soc*., 151(2004) D113.
[13] D. W. Hatchett, M. Josowicz, J. Janata, and D. R. Baer, *Chem. Mater*., 11 (1999) 2989.
[14] B. J. Hwang, R. Santhanam, and Y.L. Lin, *Electroanalysis,* 15 (2003) 1667.
[15] J. A. Smith, M. Josowicz, and J. Janata, *Phys. Chem. Chem. Phys*., 7 (2005)3614.
[16] J. A. Smith, M. Josowicz, M. Engelhard, D.R. Baer, and J. Janata, *Phys. Chem. Chem. Phys*., 7 (2005) 3619.
[17] Z. Wang, J. Yuan, M. Li, D. Han, Y. Zhang, Y. Shen, L. Niu, and A. Ivaska, *J. Electroanal. Chem*., 599(2007) 121.
[18] J. Li and X. Lin, *Sensors and Actuators*. B, 126 (2007) 527.
[19] Yassar, J. Roncali, and F. Garnier, *J. Electroanal. Chem*., 255 (1988) 53.

[20] I. M. F. De Oliveira, J. C. Moutet, and S. Hamar-Thibault, *J. Mater. Chem.,* 2 (1992) 167.

[21] Leone, W. Marino, and B. Scharifker, J. Electrochem. Soc., 139 (1992) 438.

[22] L. Torsi, M. Pezzuto, P. Siciliano, R. Rella, L. Sabbatini, L. Valli, and P. G. Zambonin, *Sensors and Actuators* B, 48 (1998) 362.

[23] N. Cioffi, L. Torsi, L. Sabbatini, P.G. Zambonin, and T. Bleve-Zacheo, *J. Electroanal. Chem.,* 488 (2000) 42.

[24] V. Tsakova, S. Winkels, and J.W. Schultze, *J. Electroanal. Chem.*, 500 (2001) 574.

[25] Yu. M. Maksimov, E. A. Kolyadko, A. B. Shishilova, and B. I. Podlovchenko, *Russ. J. Electrochem*., 37 (2001) 777.

[26] K.-M. Mangold, F. Meik, and K. Juttner, *Synth. Met*., 144 (2004) 221.

[27] Frydrychewicz, S.Y. Vassilev, G.A. Tsirlina, and K. Jackowska, *Electrochim. Acta*, 50 (2005) 1885.

[28] Mourato, S.M. Wong, H. Siegenthaler, and L.M. Abrantes, *J. Solid State Electrochem.,* 10 (2006)140.

[29] Mourato, J.P. Correia, H. Siegenthaler, and L.M. Abrantes, *Electrochim.* Acta, 53 (2007) 664.

[30] M. Ilieva, V. Tsakova, and W. Erfurth, *Electrochim. Acta*, 52 (2006)816.

[31] M. Grzeszczuk and P. Poks, *Electrochim. Acta*, 45 (2000)4171.

[32] Nirmala Grace and K. Pandian, *Electrochem. Commun*., 8, (2006)1340.

[33] K. Bouzek, K.M. Mangold, and K. Juttner, *Electrochim. Acta*, 46 (2000)661.

[34] K. Juttner, K.M. Mangold, M. Lange, and K. Bouzek, *Russian J. Electrochem.,* 40 (2004) 317.

[35] P.O. Esteban, J.M. Leger, C. Lamy, and E. Genies, *J. Appl. Electrochem*., 19 (1989) 462.

[36] R. Kostecki, M. Ulmann, J. Augustynski, D.J. Strike, and M. Koudelka-Hep, *J. Phys. Chem*., 97 (1993)8113.

[37] C. T. Hable and M.S. Wrighton, *Langmuir,* 9 (1993) 3284.

[38] E. K. W. Lai, P. D. Beattie, and S. Holdcroft, *Synth. Met.,* 84 (1997)87.

[39] F. Ficicioglu and F. Kadirgan, *J. Electroanal. Chem*., 451(1998) 95.

[40] M. J. Croissant, T. Napporn, J.M. Leger, and C. Lamy, *Electrochim. Acta.*, 43 (1998)2447.

[41] Coutanceau, M.J. Croissant, T. Napporn, and C. Lamy, *Electrochim. Acta.,* 46 (2000)579.

[42] C. H. Yang and T. C. Wen, *Electrochim. Acta*, 44 (1998)207.

[43] Kelaidopoulou, E. Abelidou, A. Papoutsis, E.K. Polychromiadis, and G. Kokkinidis, *J. Appl. Electrochem*., 28 (1998) 1101.

[44] R. Schrebler, M.A. Del Valle, H. Gomez, C. Veas, and R. Cordova, *J. Electroanal. Chem*., 380 (1995) 219.

[45] M. Malik, M. Galkowski, H. Bala, B. Grzybowska, and P. Kulesza, Electrochim. Acta, 44 (1999) 2157.

[46] A. M. Castro-Luna, *J. Appl. Electrochem*., 30 (2000)1137.

[47] A. A. Mikhaylova, E. B. Molodkina, O. A. Khazova, and V. S. Bagotzky, *J. Electroanal. Chem*., 509 (2001) 119.

[48] Lima, C. Coutanceau, J. M. Leger, and C. Lamy, *J. Appl. Electrochem*., 31 (2001)379.

[49] Lima, F. Hahn, and J.M. Leger, *Russ. J. Electrochem*., 40 (2004)369.

[50] C.C. Hu, E. Chen, and J. Y. Lin, *Electrochim. Acta*, 47 (2002)2741.

[51] E. C. Venancio, W. T. Napporn, and A. Motheo, *Electrochim. Acta,* 47(2002)1495.

[52] T. Kessler and A.M. Castro Luna, *J. Appl. Electrochem*., 32 (2002)825.

[53] T. Kessler and A.M. Castro Luna, *J. Solid State Electrochem*., 7 (2003) 593.

[54] L. Niu, Q. Li, F. Wei, X. Chen, and H. Wang, Synth. Met., 139 (2003) 271.

[55] L. Niu, Q. Li, F. Wei, X. Chen, and H. Wang, *J. Electroanal. Chem*., 544 (2003)121.

[56] L. H. Mascaro, D. Goncalves, and L.O.S. Bulhoes, *Thin Solid Films*, 461 (2004) 243.

[57] G. Wu, L. Li, J.H. Li, and B.Q. Xu, *Carbon,* 43 (2005) 2579.

[58] G. Wu, L. Li, J.H. Li, and B.Q. Xu, *J. Power Sources,* 155 (2006)118.

[59] D.W. Hatchett, R. Wijeratne, and J.M. Kinyanjui, *J. Electroanal. Chem.,* 593 (2006)203.

[60] Z. Tang, D. Geng, and G. Lu, Thin Solid Films, 497 (2006) 309.

[61] L. M. Huang, W. R. Tang, and T.C. Wen, J. Power Sources, 164 (2007) 519.

[62] S. Palmero, A. Colina, E. Munoz, A. Heras, V. Ruiz, and J. Lopez-Palacios, *Electrochem. Commun.,* 11 (2009)122.

[63] S. Dominguez-Dominguez, J. Arias-Pardilla, A. Berenguer-Murcia, E. Murallon, and D. Cazorla-Amoros, *J. Appl. Electrochem*., 38 (2008)259.

[64] F.J. Liu, L.M. Huang, T.C. Wen, C.F. Li, S.L. Huang, and A. Gopalan, *Synth. Met*., 158 (2008)603.

[65] F.J. Liu, L.M. Huang, T.C. Wen, C.F. Li, S.L. Huang, and A. Gopalan, *Synth. Met.,* 158 (2008)767.

[66] F. T. A. Vork, L. J. J. Janssen, and E. Barendrecht, *Electrochim. Acta*, 31 (1986) 1569.

[67] S. Holdcroft, and B. L. Funt, *J. Electroanal. Chem.*, 240 (1988) 89.

[68] D. J. Strike, N. F. De Rooij, M. Koudelka-Hep, M. Ulmann, and J. Augustynski, *J. Appl. Electrochem.*, 22 (1992) 922.

[69] M. Hepel, *J. Electrochem. Soc.*, 145.(1998) 124.

[70] C.W. Kuo, L.M. Huang, T.C. Wen, and A. Gopalan, *J. Power Sources,* 160 (2006) 65.

[71] J.Y. Lee and T.C. Tan, *J. Electrochem. Soc.*, 137 (1990)1402.

[72] V. Tsakova and A. Milchev, *Electrochim. Acta*, 36 (1991)1151.

[73] N. Hernandez, J.M. Ortega, M. Choy, and R. Ortiz, *J. Electroanal. Chem.,* 515 (2001)123.

[74] S. Ivanov and V. Tsakova, *Electrochim. Acta*, 49 (2004) 913.

[75] S. Ivanov and V. Tsakova, *Electrochim. Acta*, 50 (2005) 5616.

[76] N. Cioffi, L. Torsi, I. Losito, C. Di Franco, I. De Bari, L. Chiavarone, G. Scamarcio, V. Tsakova, L. Sabbatini, and P.G. Zambonin, *J. Mater. Chem.,* 11 (2001) 1434.

[77] V. Tsakova, D. Borissov, and S. Ivanov, *Electrochem. Commun.*, 3 (2001)312.

[78] V. Tsakova, D. Borissov, B. Ranguelov, Ch. Stromberg, and J. W. Schultze, *Electrochim. Acta*, 46 (2001)4213.

[79] V. D. Jovic, T. Trisovic, B.M. Jovic, and M. Vojnovic, *J. Electroanal. Chem.*, 408 (1996) 149.

[80] S. S. Kumar, J. Mathiyarasu, and K. L. Phani, *J. Electroanal. Chem.,* 578 (2005) 95.

[81] J. Mathiyarasu, S. Senthilkumar, K.L.N. Phani, and V. Yegnaraman, *Mater. Lett.*, 62 (2008)571.

[82] Zanardi, F. Terzi, L. Pigani, A. Heras, A. Collina, J. Lopez-Pallacios, and R. Seeber, *Electrochim. Acta,* 53 (2008)3916.

[83] F. Terzi, C. Zanardi, V. Martina, L. Pigani, and R. Seeber, *J. Electroanal. Chem.,* 619 (2008)75.

[84] M. V. Henry, C. C. Hsueh, B. P. Timko, and M.S. Freund, *J. Electrochem. Soc.*, 148 (2001)D155.

[85] E. T. Kang, Y.P. Ting, KG. Neoh, and K.L Tan, *Polymer*, 34 (1993) 4494.

[86] E. T. Kang, Y.P. Ting, K.G. Neoh, and K.L. Tan, *Synth. Met.,* 69(1995)477.

[87] K. G. Neoh, T.T. Young, N.T. Looi, and E.T. Kang, *Chem. Mater.*, 9 (1997) 2906.

[88] M. S. A. Abdou and S. Holdcroft, *Synth. Met.*, 60 (1993) 93.
[89] M. S. A. Abdou and S. Holdcroft, *Chem. Mater.*, 8(1996)26.
[90] J. Wang, K. G. Neoh, and E. T. Kang, *J. Colloid Interface Sci.*, 239 (2001) 78.
[91] M. A. Khan, C. Perruchot, S.P. Armes, and D.P. Randall, *J. Mater. Chem.*, 11 (2001)2631.
[92] V. V. Kondratiev, N.A. Pogulaichenko, E.G. Tolstopjatova, V. V. Malev, *J. Solid State Electrochemistry*, 15 (2011) 2383.
[93] V. V. Kondratiev, N. A. Pogulaichenko, SuoHui, E. G. Tolstopjatova, V. V. Malev, *J. Solid State Electrochemistry*, 16 (2012)1291.
[94] L.M. Abrantes and J.P. Correia, *Mater. Sci. Forum*, 191 (1995) 235.
[95] M. Hasik, A. Derlinkiewicz, M. Choczynski, S. Quillard, and A. Pron, *Synth. Met.*, 84 (1997)93.
[96] Derlinkiewicz, M. Hasik, and M. Choczynski, Mater. *Res. Bull.*, 33 (1998) 739.
[97] S. W. Huang, K. G. Neoh, C. W. Shih, D. S. Lim, E. T. Kang, H. S. Han, and K. L. Tan, *Synth. Met.*, 96 (1998)117.
[98] M. Josowicz, H. S. Li, K. Domansky, and D. R. Baer, *Electroanalysis,* 11 (1999)774.
[99] Derlinkiewicz, M. Hasik, and M. Kloc, *Synth. Met.*, 102 (1999) 1307.
[100] Z. H. Ma, K. L. Tan, and E. T. Kang, *Synth. Met.*, 114 (2000) 17.
[101] V.W. L. Lim, E.T. Kang, and K. G. N*eoh, Synth. Met.*, 123 (2001)107.
[102] M. Hasik, A. Derlinkiewicz, and E. *Wenda, Synth. Met.*, 119(2001) 335.
[103] M. Hasik, E. Wenda, C. Paluskiewicz, A. Bernasik, and J. Camra, Synth. Met., 143(2004) 341.
[104] Derlinkiewicz, M. Hasik, J. W. Sobczak, E. Sobczak, A. Bernasik, and E. Bielanska, *Mater. Res. Bull*, 40 (2005)869.
[105] Mourato, A.S. Viana, J.P. Correia, H. Siegenthaler, and L.M. Abrantes, *Electrochim. Acta,* 49 (2004) 2249.
[106] S. N. Eliseeva, V. V. Malev, V. V. Kondratiev, *Russian J. Electrochemistry,* 45 (2009) 1045.
[107] V. V. Kondratiev, T. A. Babkova, E. G. Tolstopjatova, *J. Solid State Electrochemistry,* 17 (2013) 1621.
[108] V. V. Kondratiev, T. A. Babkova, S. N. Eliseeva *Russ. J. Electrochem.*, 48 (2012) 205.
[109] A.Q. Zhang, C.Q. Cui, J.Y. Lee, and F.C. Loh, *J. Electrochem. Soc.*, 142 (1995) 1097.
[110] M. Y. Lebedev, M. V. Lauritzen, A. E. Curzon, and S. Holdcroft, *Chem. Mater.,* 10(1998) 156.

[111] F. Y. Song and K.K. Shiu, *J. Electroanal. Chem.*, 498 (2001)161.
[112] Visy, E. Pinter, T. Fullei, and R. Patakfalvi, *Synth. Met.*, 152 (2005)13.
[113] E. Pinter, R. Patakfalvi, T. Fiilei, Z. Gingl, I. Dekany, and C. Visy, *J. Phys. Chem.* B, 109 (2005) 17474.
[114] V. M. Jovanovic, S. Terzic, and A. Dekanski, J. *Serb. Chem. Soc.*, 70 (2005)41.
[115] R. Dimeska, P.S. Murray, S.F. Ralph, and G.G. Wallace, *Polymer,* 47 (2006) 4520.
[116] M. Ocypa, M. Ptasinska, A. Michalska, K. Maksymiuk, and E.A.H. Hall, *J. Electroanal. Chem.*, 596 (2006)157.
[117] E. Pinter, Z.A. Fekete, O. Berkesi, P. Makre, A. Patzko, and C. Visy, *J. Phys. Chem.* C, 111 (2007)11872.
[118] J. Stejskal, J. Prokes, and I. Sapurina, *Mater. Lett.,* 63(2009) 709.
[119] E. G. Tolstopyatova, N. A. Pogulyaichenko, and V. V. Kondratiev, *Russian J. Electrochemistry,* 50 (2014) 510.
[120] R Zeis, T Lei, K Sieradzki, J Snyder, J Erlebacher *J. Catal*, 253 (2008)132.
[121] V. V. Malev, O. V. Levin, *Electrochimica Acta*, 56 (2011) 3856.
[122] V. V. Malev, O. V. Levin, V.V. Kondratiev, *Electrochimica Acta,*122 (2014) 234.
[123] V. G. Levich, Physico-Chemical Hydrodynamics, Moscow: *Phys. Math. Giz.,* 1959 [in Russian].
[124] S. Chakraborty, C. Retna Raj, *Sensors and Actuators* B, 147(2010) 222.

6. Charge Transfer in the Films Formed by Polymeric Complexes of Nickel with the Schiff Base Ligands

Metal-containing polymer films can be prepared not only as heterogeneous structures, but also due to incorporation of chemically bonded metal atoms into the organic backbone of a polymer. Here we would like to summarize the results received at studying of one of such compounds, namely polymer films of Ni(Schiff) complexes (figure 6.1).

Electrode modification by anodically polymerized films of metal N, N′-bis(salicylidene)ethylenediamine (salen) complexes was first demonstrated in 1988 [1, 2]. Horwitz and Murray [1]polymerized the 4,4'-di(N-methyl-N-phenylmethyl)salen complexes of nickel(II), cobalt(II), and manganese(III) onto the surfaces of platinum electrodes. Independently, electropolymerization of nickel(II) salen onto both the platinum and tin oxide electrodes was performed by Goldsby and coworkers [2]. Almost simultaneously Timonov and Shagisultanova have synthesized salen-type polymeric complexes of palladium [3].

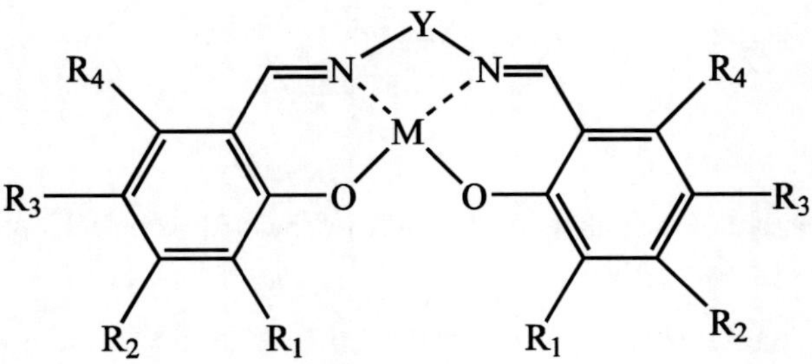

Figure 6.1. General formula of the metal complexes with salen-type ligand; M- metal atom, R_1-R_3 – hydrogen or organic radical, Y – bridge group.

Literature data [4] give a list of conditions that are important for formation of conducting polymer films by electropolymerization of metal-salen complexes:

1. ability of the metal center (M in Figure 6.1) to have at least two different oxidation states (+2,+3,+4);
2. existence of π-conjugated network in the ligand;
3. planar geometry of monomer complex;
4. low coordination ability of the solvent, from which polymer is formed;
5. oxidation potential must be high enough to ensure two-electron oxidation of the monomer, but low enough not to destroy the π-conjugated network at overoxidation of the complexes (typical values of potential are 0.9 – 1.3 V vs. Ag/AgCl (sat.)).

The existence of a ligand-based π-conjugated network together with electrochemically active central metal atom makes it possible to consider the polymeric complexes as a unique type of CPs. Mechanisms of their electropolymerization and charge transfer have been the subjects of discussions for years.

In general, an oxidative electropolymerization of salen-type metal complexes is described as an oxidative coupling of cation-radicals with formation of C-C linkage, leading to formation of polyphenylene-type polymers [2, 5, 6] (figure 6.2).

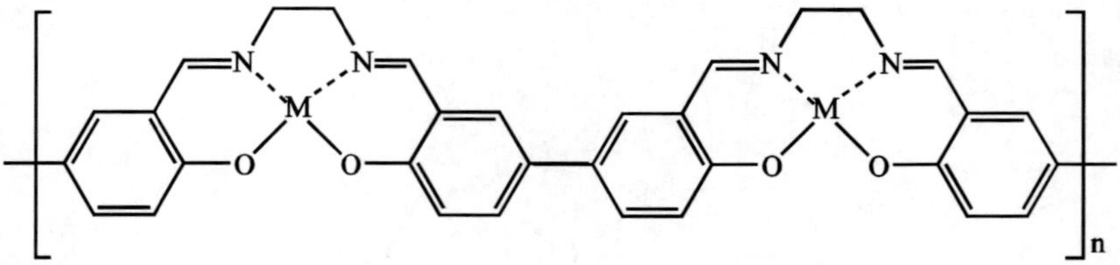

Figure 6.2. Structure of the polymeric complex [M(Salen)] according to the oxidative coupling polymerization model.

Such polymerization route is confirmed by chemical analysis of polymer decomposition products [7] and impossibility of electropolymerization of 5-alkyl substituted complexes [2]. However, some experimental facts could not be explained in the scope of this model, namely:

- electropolymerization of 5-Br substituted complexes [8];
- absence of electropolymerization for complexes with non-planar structure [3];
- low conductivity of the Zn polymeric complexes, where the central atom can have only one (2+) oxidation state [7];
- conductivity of electropolymerized films in the direction normal to the electrode surface is 6-9 orders higher than that of in directions tangential to the electrode surface [9].

It is well known that a solid phase of planar monomeric complexes of transition metals (including salen-type complexes) consists of dimeric aggregates or linear chains, linked by d^8-d^8interaction between metal atoms [10-12]. This fact together with the above observations has led to the appearance of another polymer formation model, which assumes formation of a stack-type supramolecular structure with intermolecular charge transfer [13] (figure 6.3).

Now the most plausible model of polymers molecular structure is the hybrid one proposed by Piters et. al. [14] and confirmed by other groups [8, 15, 16]. According to the model, polymer films consist of fragments linked to each other by both C-C covalent bonds and intermolecular interactions (figure 6.4).

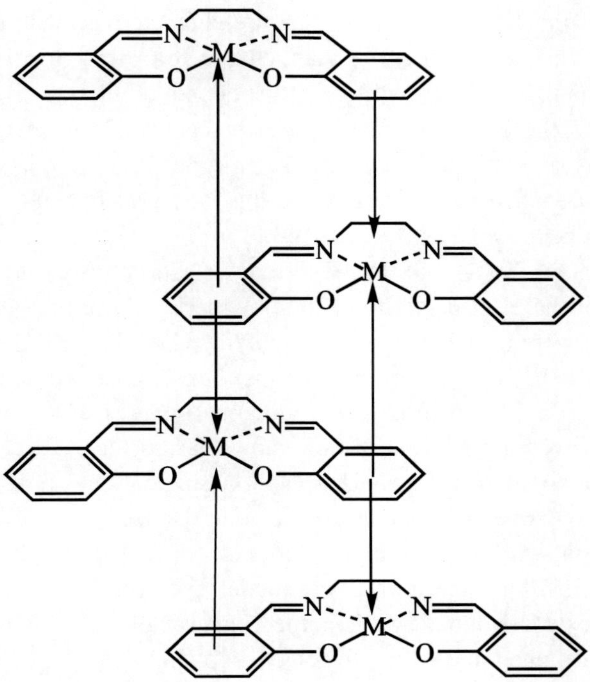

Figure 6.3. Stack model of molecular materials based on salen-type complexes of the transition metals. Arrows indicate the routes of intermolecular charge transfer.

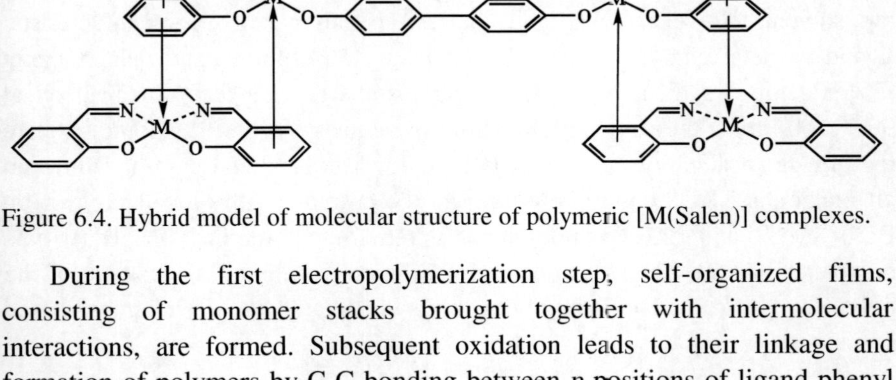

Figure 6.4. Hybrid model of molecular structure of polymeric [M(Salen)] complexes.

During the first electropolymerization step, self-organized films, consisting of monomer stacks brought together with intermolecular interactions, are formed. Subsequent oxidation leads to their linkage and formation of polymers by C-C bonding between *p*-positions of ligand phenyl rings. As a result, the obtained polymers are characterized with the capability of intermolecular and intramolecular charge transfer proceeding due to different mechanisms and providing different conductivities.

As well as the electropolymerization mechanism, the mechanism of the redox transformations in the polymeric salen-type complexes is still not clear. Here, we will demonstrate the recent findings in studies of such transfer on the example of nickel complexes.

Initially, the Ni(II) -> Ni(III) transformation was assumed to be responsible for the electroactivity of the polymeric [Ni(Schiff)] complexes, so such materials were designated as redox polymers[14, 17, 18]. However, direct detections of Ni(III) by electron paramagnetic resonance methods have failed [6, 19-21]. Indeed, studies of oxidized monomeric SalEn-type complexes with substituents preventing polymerization show that, in the oxidized state, these complexes can be described as Ni(II)-phenoxyl radical ones instead of Ni(III)-phenolate, if no exogenous ligands are added to the used solution [22-25]. This was the reason why ligand-based model of polymers electroactivity was suggested [6, 19-21]. According to the model, the polymers were assumed to have the polaron conductivity like other intrinsically conducting polymers. However, some phenomena could not be explained in scope of this model [26]. It is known that the electronic structure of oxidized Ni(II) square-planar complexes is sensitive to many factors. Electron paramagnetic resonance (EPR) spectroscopy data prove that exogenous ligands, such as pyridine or highly coordinating solvents (e.g. dimethylformamide) are localized in the axial positions and, thus, stabilize Ni(III) oxidation state [2, 27-29].

The same might happen in polymeric nickel SalEn-type complexes, and a high oxidation state of nickel atoms becomes stable through coordination of solvent molecules in axial positions. In such case, *ex situ* detection of Ni(III) will be impossible in result of the Ni(III)/Ni(II) transformation after removing the solvent. However, Ni(III) in SalEn-type complexes has characteristic ligand-to-metal charge transfer (LMCT) light absorption band near 500 nm [25]. Phenoxyl radical has ligand-to-ligand charge transfer (LLCT) band at 900-1100 nm in the same spectra. So, the intensity of the LLCT indicates to the charge localization degree; it is high for a delocalized system, while no characteristic LLCT band is observed in the case of a fully localized system. Both LMCT and LLCT bands are observed for polymeric [Ni(SalEn)] type complexes [6, 19, 21, 30], so one can assume a combined mechanism of the polymers electrooxidation involving both the ligand based and metal based processes.

Cyclic voltammetry data confirms such complicated route of redox transformations in the polymeric complexes in question. It will be demonstrated on examples of a series of nickel complexes having different ligand structures (figure 6.5).

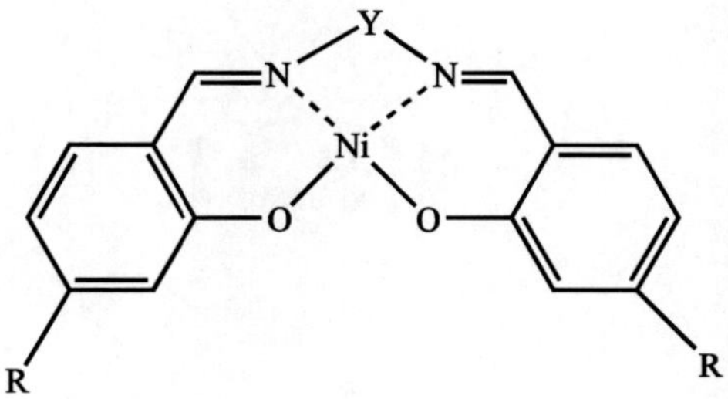

Figure 6.5. Monomer complexes. Abbreviations:

R=*H*, Y=CH_2-CH_2 – [Ni(SalEn)]

R= OCH_3, Y= CH_2-CH_2 – [Ni(CH_3OSalEn)]

R= *H*, Y= $C(CH_3)_2$-$C(CH_3)_2$ – [Ni(SaltmEn)]

R=OCH_3, Y= $C(CH_3)_2$-$C(CH_3)_2$ – [Ni(CH_3OSaltmEn)]

R= *H*, Y=  – [Ni (SalPhen)]

R= OCH_3, Y=  – [Ni (CH_3OSalPhen)]

Cyclic voltammograms of unsubstituted poly[Ni(SalEn)] have a pair of broad peaks (Figure 6.6a). Polymeric complexes with substitued ligands have different voltammetric responses. Such responses of complexes with methyl substituents in imino bridges have shoulders on both the anodic and cathodic branches (poly[Ni(SaltmEn)], Figure 6.6b). Introduction of the methoxy substituents in the phenyl rings leads to the appearance of additional voltammetric peaks together with the observed shift of polymer electroactivity in the cathodic direction (poly[Ni(CH3OSalEn)], poly[Ni(CH3OSaltmEn)]], Figure 6.6c,d). Two pairs of voltammetric peaks are usually interpreted as an evidence of two electrochemical processes.

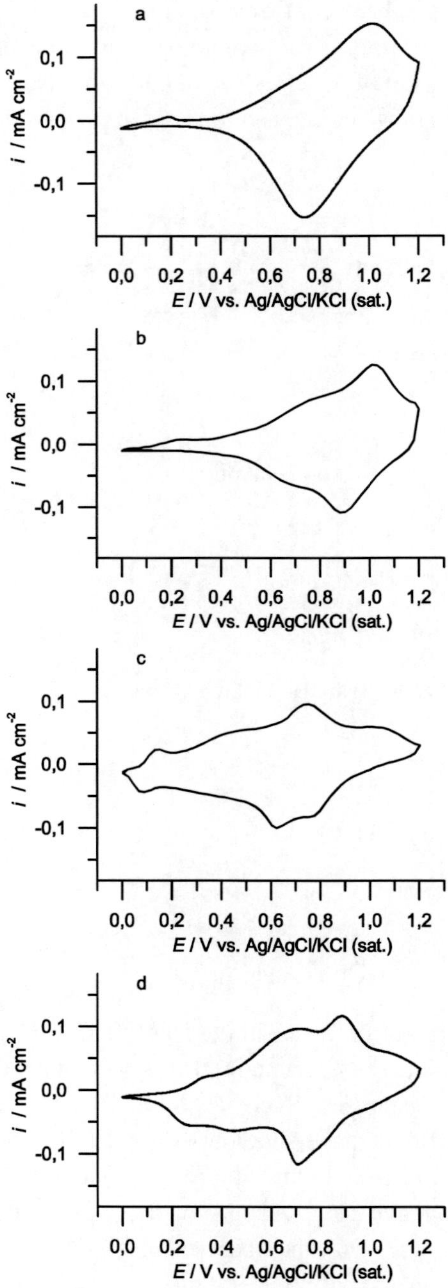

Figure 6.6. Typical voltammograms of polymeric complexes, a) poly[Ni(SalEn)], b) poly[Ni(SaltmEn)], c) poly[Ni(CH_3OSalEn)], d) poly[Ni(CH_3OSaltmEn)].

In the literature, it was attributed to structurally determined transport [19, 30] or to some combination of the ligand-based and metal-based oxidation processes [31, 32]. It can also be noticed that the reduced currents (i.e. the ratios of I/L, where L, the film thickness) recorded for two different values of L (approximately equal to 1 and 2 μm) at scan rate ν = 50 mV/s differ insignificantly from each other, demonstrating only a small shift of the peak potentials. This allows one to assume a quasi-monolayer occupancy of both films with their ox- and red-fragments in the course of recording CV-curves, i.e. the practical absence of charge carriers' gradients within the film interior.

Impedance spectra of the polymer nickel complexes are also quite unusual [33]. Typical Nyquist plots (see figure 6.7) are characterized with the absence of high frequency semicircles (at least, at frequencies lower than 100 kHz) and hardly resolvable Warburg and pseudo-capacitive changes in the imaginary (–Z") and real (Z') impedance constituents with decreasing angular frequency $\omega = 2\pi f$ (10^5 Hz > f > 10^{-1} Hz) of the applied alternating potential (its amplitude is 10 mV) [31]. At potentials on the edges of the film electroactivity region (300÷400 and 1000 mV) and *ac* frequencies below 1 kHz, the Warburg constituent of the impedance is explicitly revealed as linear dependences of Z'',Z' in functions of the reversal square rooted frequency with the slope close to π/4 [34]. Thus, the obtained data of impedance measurements appear to be non-trivial most likely resulting from several redox transitions.

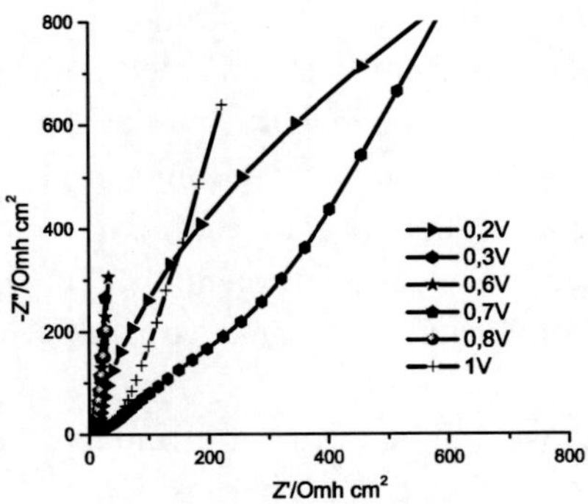

Figure 6.7. Impedance spectra of 1 μm thick poly-[Ni(CH_3O- SalphEn)] film on 0.07 cm^2 glassy carbon electrode in solution of 0.1 M Et_4NBF_4 in acetonitrile. Potentials of the spectra registrations indicated on the graph are given vs. Ag/AgCl/KCl (sat.).

The impedance theory is well developed for polymer films containing two kinds of charge carriers [34]. However, the systems in question do not satisfy the model assumptions of such theory due to the presence of several redox transitions, the potential regions of which overlap each other. It appeared that the impedance $Z(\omega)$ in the regions of overlapping redox transitions can be represented as follows[31]:

$$Z(\omega) = Z'(\infty) + (1-j)\sigma_W/\omega^{1/2} - j/\omega C_{LF}, \qquad (6.1)$$

where $Z'(\infty)$ is the value of the impedance extrapolated at $f = \infty$, $j = \sqrt{-1}$; angular frequency $\omega = 2\pi f$; C_{LF} is the low-frequency capacity of the film; σ_W, the Warburg constant. Such form of the frequency dependent impedance, which is not predicted by the existing theory [34], is a combination of the Warburg and pseudo-capacitive responses. It seems that the necessary condition for observation of such impedance response is the existence of two overlapping redox transitions, which should be accompanied with some cross effects resulting from the presence of three kinds of charge carriers in the film interior instead of two ones assumed in the existing theory.

To estimate what a factor is exactly responsible for the complicated voltammetric and impedance responses of the polymeric complexes, it is first necessary to establish stoichiometric relationships of the charge transfer process investigated. Number ***n*** of electrons consumed for reduction/oxidation of one repeat unit of the polymers was determined by parallel CV and EQCM measurements in [35]. In particular, it was found that number ***n*** is equal to 1.0 ± 0.1 for all the poly-[Ni(Schiff)] films. This means that the polymer fragment oxidation/reduction is exactly one electron process. It is just the reason why one needs to assume the existence of three kinds of charge carriers within the polymers, in order to explain the appearance of two (or even more) transitions separated sufficiently in the electrode potentials. Though the nature of such charge carriers is still questionable, some verisimilar speculations on their entity seem possible.

As mentioned above, three structural models are discussed in the literature, to explain the existence of such transitions:

- as typical conjugated chains with polaron and bipolaron conductivity [6, 7],
- as stacked supramolecular structures with metal-based conductivity [18, 36],

– as some combination of the previous ones [6, 14, 37].

The appearance of two redox transitions in scope of each model was discussed on the basis of thermodynamic considerations [32]. The authors considered three particular electrode systems corresponding to the assumed existence of three kinds of charge carriers within a polymer film. The simplest system corresponds to a film containing two kinds of immobile carriers that are not capable to transform into each other ("mechanical mixture of two immobile charge carriers"). The second analyzed system corresponds to a polymer film, in which the polaron/bipolaron conversion takes place:

$$2R - e \leftrightarrow P \text{ and } P - e \leftrightarrow B \tag{6.2}$$

The third model system describes polymer films possessing both the polaron and redox conductance.

Usually, the polaron charge is distributed over 3÷6 neighboring fragments of polymer chains [4]. In case of complexes with the Schiff bases, phenol rings can be considered as possible sites for such distribution, so that two monomer units might provide four sites for the polaron localization. For the sake of simplicity, we therefore assumed that polarons P and bipolarons B are delocalized over 2 polymer repeat units R; this is the reason, why the polaron formation is described by the first reaction from reactions (6.2). At the same time, the redox reaction of changing the electron state of nickel atoms

$$R - e \leftrightarrow Ox \tag{6.3}$$

is also possible (here, Ox symbolizes a [Ni(III)Schiff]-unit). The above reactions, in essence, mean the existence of the third reaction proceeding in the film interior, namely the following one:

$$P \leftrightarrow Ox + R, \tag{6.4}$$

which can be treated as a disproportionation reaction, i.e. the electron exchange reaction that might proceed in any polaron with formation of the reduced and oxidized states of the nickel atoms included into its structure.

After solving the corresponding thermodynamics equations, it was found that, in all the above cases, splitting of voltammetric peaks into two apparent transitions can be observed at proper ratios of the constants characterizing the reactions assumed [32]. However, the obtained results are not sufficient to determine the real route of redox transformations by basing exclusively on the obtained voltammetric curves. To determine such route, one should refer to the data obtained by other techniques. The obtained electrochemical quartz crystal microbalance (EQCM) data demonstrate that oxidation of the polymers in question is accompanied with an increase in their weight $\Delta m(Q)$ due to entering charge-compensating ions and solvent molecules into the film, whereas the reversal process takes place under the polymer reduction. However, the molar weights of charge-transferring species occur to be dependent on the oxidation potential [20]. One can find two linear regions on the obtained $\Delta m(Q)$-curves, where Q, the consumed charge. At the oxidation potentials below 0.8 V vs. Ag/AgCl (sat.), the molar masses of transferring species are less in all cases than the mass of BF_4^--anion, which corresponds to replacement of a solvent molecule by each charge compensating anion. If oxidation is continued up to the electrode potentials of about 1.2 V, no solvent ejection is observed; the mass of charge-compensating species becomes higher than that of BF_4^--anion, so the flux of counter-ions is most probably accompanied by solvent injection into the film. The latter indicates to some expansion of the film under these conditions. These peculiarities were confirmed by probe-beam study [20] and *in-situ* ellipsometric study of Pd[CH_3OSalEn] films [26]. From the above voltammetric, impedance, and EQCM results it becomes clear that the polymer oxidation proceeds by two steps, and the second step is accompanied by solvent injection into the film.

As concerns the nature of the observed redox transformations, it might be uncovered with the help of spectral methods. Characteristic UV-visible spectra of poly[Ni(SalEn)] films measured at their potentiostatic polarization from 0 V to 1.2 V (vs. Ag/AgCl (sat.)) are presented in Figure 6.8. The reduced form of the polymer (Red) has a characteristic absorption band at 406 nm and another one in near-UV range, which is not well resolved due to some system limitations. In result of the polymer oxidation new absorption bands appear. Namely up to 0.9 V, a growth of the wide near-infrared band is observed. Such low energy band is characteristic for delocalized charge state, which can be identified as a polaron in terms of the conducting polymer theory (similar bands are observed for PEDOT films, see the preceding section).

However at potentials higher than 0.9 V, some decrease of the indicated absorption band is observed together with an increase of absorbancy at 460 nm, leading to the formation of a well-defined band at polarizations higher than 1 V. According to the data obtained for monomeric complexes [27-29], the appearance of such band corresponds to localization of charge density on Ni atoms in the presence of axial ligands.

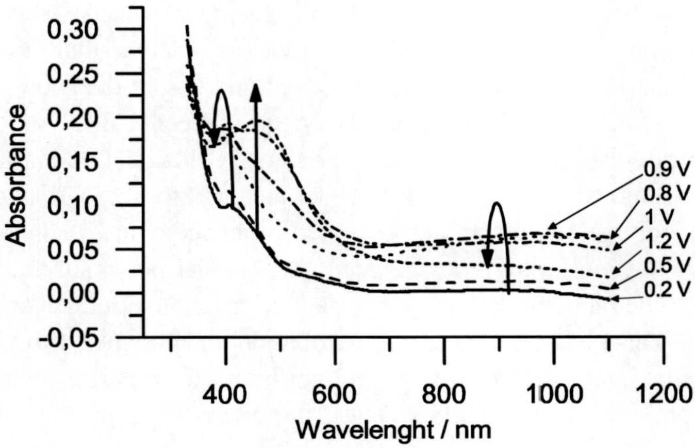

Figure 6.8. Characteristic spectra of poly[Ni(SalEn)] films, measured at potentiostatic polarization from 0 V to 1.2 V. Directions of absorption maxima evolution with an increase of the electrode potential are indicated by arrows.

In this case, we should assume that the highly oxidized state of the polymer is metal-centered and the axial positions of Ni atoms are occupied by some exogenous ligands.

We assume that moderately coordinating solvents, such as acetonitrile may act as axial ligands in case of the studied polymer films. If the oxidized film is removed from the working cell and dried under vacuum without exposure to the air, a rapid decrease of the band at 460 nm is observed together with increasing the near-infrared band. This demonstrates a crucial role of a coordinating solvent in the stabilization of the highly oxidized form of the polymers in question.

Here, we would like to conclude that the oxidation of SalEn-type polymeric nickel complexes proceeds in two steps. The reduced form of the polymer is first oxidized to the form, which is characterized by charge delocalization over the conjugated bonds system of the polymer. Subsequent oxidation in acetonitrile solutions leads to formation of the second oxidized form, spectra of which cannot be described in terms of delocalized charge, but, in our opinion, correlate with charge localization on metal atoms stabilized by axial coordination of solvent molecules or another particles.

Such coordination may cause important features of the polymers if the axial ligands are also electroactive in the range of polymer's electroactivity. In non-aqueous solutions at the water traces being carefully removed, the most of SalEn-type polymeric metal complexes are quite stable if cycled in the potential range below 1.2 V. Even if a small amount of water is added to the bathing solution this leads to a film degradation (i.e. gradual loss of its electroactivity), the rate of which increases with increasing the anodic limit of the potential scan. If such anodic limits do not exceed 0.8 V, the most of [Ni(Schiff)] polymers are stable even after some prolonged cycling in water-containing solutions. However, attempts to increase the oxidation potential lead to the polymer degradation. In mixed aqua-acetonitrile solutions, the water decomposition is thermodynamically allowed at potentials higher than 0.8 V, but its rate on the bare substrate is negligible due to strong adsorption of acetonitrile molecules. In this case, defoliation of polymer films by gas bubbles produced on the substrate should not be responsible for the observed loss of the polymers electroactivity. Thus, the water decomposition might be responsible for the film degradation if and only if it is somehow catalyzed by polymer fragments. To determine how water interacts with polymer structure, we have performed a comparative study of the polymer films stability with different substituents in the ligands of their monomer precursors.

Introduction of the metoxy substituents in the third positions of phenyl rings increases the elecrtronic density in the aldehyde moieties. It shifts the beginning of the polymer electroactivity range by more than 200 mV in the cathodic direction.

Phenyl substituents in the imine bridge decrease the electronic density mostly on nitrogen and metal atoms through the system of conjugated bonds and shift the electroactivity region to the anodic direction. If water molecules attack any kind of aldehyde atoms, one may expect a strong influence of the metoxy substituents on the stability of Salen-type polymeric complexes of nickel. If bridge atoms (for example, imine nitrogens) are the main targets for water attack, then the influence of phenyl substituents on the stability is expected. Replacement of nickel by other metals should influence the stability of complexes provided that water molecules interact with metal atoms. We have performed stability tests for a series of substituted NiSalen type polymers by cyclic voltammetry in acetonitrile solution of 0.1 M $TEABF_4$ in the presence of 1 volume percent of water. In figure 6.9 the relative charge consumed for the reduction of polymers is presented as a function of the scan number.

It can be noted that the metoxy substituent has a negligible effect on the polymer degradation rate. This gives the evidence that aldehide moiety does not participate in the reaction with water. Exchange of ethyl groups with the phenyl ones in imine bridge leads to some inhibition of degradation, however the observed effect is rather small. In contrast to ligand groups, metal atoms have an extreme influence on the stability of Salen-type complexes. In particular, films formed from copper complexes also degradate if cycled in the presence of water, but palladium ones are stable.

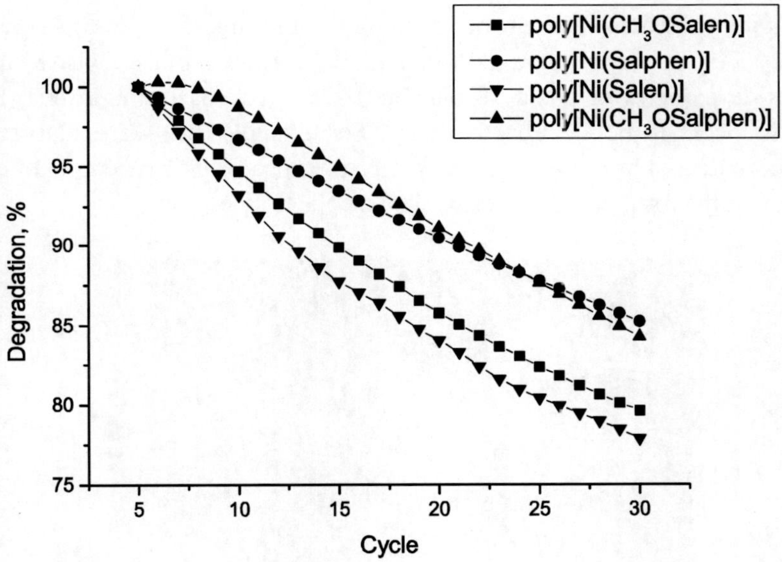

Figure 6.9. Relative charge consumed for the reduction of polymers as a function of the scan number. The represented data were obtained by integration of the cathodic branches of cyclic voltammograms. Scan rate equals 50 mV/s, potential range is of 0 ÷ 1 V vs. Ag/AgCl (sat.)

To explain such phenomena, one should remember that, as was shown for monomeric complexes of similar structure, copper and nickel complexes can be oxidized to higher oxidation states in the presence of exogenous ligands, such as pyridine or dimethilsulfoxide [2, 23, 25]. At the same time, as far as we know, no evidence of the existence of palladium atoms in the oxidation state higher than two has been ever observed for Salen-type complexes. Water has a donor number higher than acetonitrile, so it might be coordinated on axial positions. Then, as soon as the potential becomes high enough, water molecules start to decompose with the help of highly oxidized metal atoms and this process leads to cleavage of neighboring bonds or transition of metal atoms from the initial complex to a more stable hydrated form, such as $Ni(OH)_2$ or NiOOH. Formation of the later compounds can be detected after electrochemical oxydation of the film in alkaline media. As it starts, an alkaline electrolyte destroys the ligand polymer matrix and nanoparticles of the metal hydroxyde are kept on the substrate surface. Polymeric structure of the Schiff base complex ensures enough the distance between nucleation centers during the hydroxide formation, so no agglomeration of the formed nanoparticles is observed.

Figure 6.10 demonstrates the scanning electron microscope (SEM) image of the obtained layer on a polished glassy-carbon surface, where nickel hydroxide particles of about 10 nanometers size can be distinguished. Such deposits occur to be electroactive ones. For example, they can catalyze the alcohol oxidation processes. Figure 6.11 proves the catalytic oxidation of ethyl alcohol on the formed catalyst in an alkaline electrolyte.

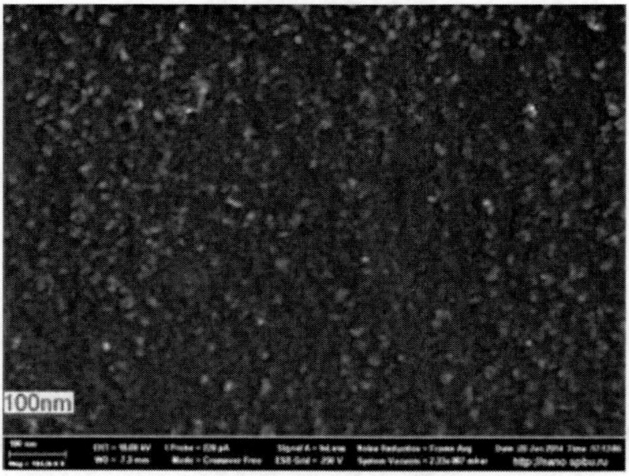

Figure 6.10. SEM image of NiOOH catalyst deposited on a glassy-carbon surface after decomposition of poly[Ni(CH_3OSalen)] film by 50 cycles in 0.2 M NaOH solution at 50 mV/s from 0 to 0.8 V vs. Ag/AgCl (sat.).

In this section, we have summarized the obtained experimental results concerning structure and charge transfer properties of polymeric complexes of nickel with Salen-type ligands. It has been demonstrated that, like other CPs discussed above, such complexes form electroactive polymer films characterized with the following features:

– joint ion and solvent transfer during redox transformations followed with reversible expansion of the films at their more profound oxidation;

– only qualitative agreement between their electrochemical properties and those predicted with the existing charge transfer theory.

Besides, such films possess unusual peculiarities in their voltammetric and impedance responses resulting from the evident existence of more than one kind of “immobile” charge carriers within the film interior.

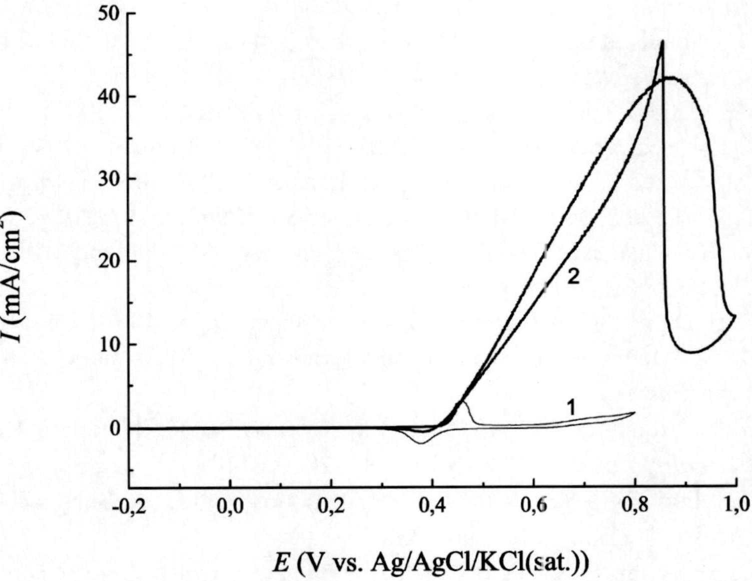

Figure 6.11. Electrooxydation of ethyl alcohol on CG electrode coated with NiOOH catalyst in a solution of 0.2 M NaOH and 1 M C_2H_5OH (curve 2). Voltammetric response of the same electrode in 0.2 M NaOH is also presented (curve 1).

References to Section 6

[1] C. P. Horwitz and R. W. Murray, *Mol. Cryst. Liq. Cryst.*, 160 (1988)389.

[2] K. A. Goldsby, J. K. Blaho, and L. A. Hoferkamp, *Polyhedron*, 8 (1989)113.

[3] I. A. Chepurnaya, P. V. Gaman’kov, T. Yu. Rodyagina, S. V. Vasil’eva, and A. M. Timonov, *Russ. J. Electrochem.*, 39 (2003)314.

[4] V. V. Malev, V. V. Kondratiev, A. M. Timonov. Polymer-modifyed electrodes. St. Petersburg: Nestir-Istoriya, 2012. –346 p.

[5] P. Audebert, P. Hapiot, P. Capdevielle, and M. Maumy, *J. Electroanal. Chem.*, 338 (1992)269.
[6] M. Vilas-Boas, C. Freire, B. De Castro, P. A. Christensen, and A. R. Hillman, *Inorg. Chem.,* 36 (1997)4919.
[7] P. Audebert, P. Capdevielle, and M. Maumy, *New J. Chem.*, 16 (1992)697.
[8] M. P. Karushev, A. M. Timonov, Izvestia: *Herzen University Journal of Humanities & Sciences*, 144 (2012)99.
[9] G. Shagisultanova, I. Orlova, L. Ardasheva, and E. Popova, *Macromolecular Symposia*, 136 (1998)91.
[10] M. A. Siegler and M. Lutz, *Crystal Growth & Design*, 9 (2009)1194.
[11] J. Fonseca, J. Martinez, L. Cunha-Silva, A. L. Magalhaes, M. Teresa Duarte, and C. Freire, *Inorg. Chim. Acta*, 363 (2010)4096.
[12] T. W. Thomas and Underhil. Ae, *Chem. Soc. Rev.*, 1 (1972)99.
[13] S. V. Vasil'eva, K. P. Balashev, and A. M. Timonov, Russ. J. Electrochem., 34 (1998)978.
[14] C. E. Dahm and D. G. Peters, *J. Electroanal. Chem.*, 406 (1996)119.
[15] B. N. Afanas'ev, Y. A. Polozhentseva, and A. M. Timonov, *Russ. J. Appl. Chem.*, 84 (2011)1341.
[16] B. N. Afanas'ev, Y. A. Polozhentseva, and A. M. Timonov, Russian *Journal of Physical Chemistry A*, 84 (2010)2148.
[17] I. E. Popeko, V. V. Vasilev, A. M. Timonov, and G. A. Shagisultanova, *Zh. Neorg. Khim.,* 35 (1990)933.
[18] C. E. Dahm, D. G. Peters, and J. Simonet, *J. Electroanal. Chem.*, 410 (1996)163.
[19] M. Vilas-Boas, I. C. Santos, M. J. Henderson, C. Freire, A. R. Hillman, and E. Vieil, *Langmuir*, 19 (2003)7460.
[20] M. Vilas-Boas, M. J. Henderson, C. Freire, A. R. Hillman, and E. Vieil, *Chemistry-a European Journal*, 6 (2000)1160.
[21] M. Vilas-Boas, C. Freire, B. De Castro, P. A. Christensen, and A. R. Hillman, *Chem. Eur. J.,* 7 (2001)139.
[22] Y. Shimazaki, T. Yajima, F. Tani, S. Karasawa, K. Fukui, Y. Naruta, and O. Yamauchi, *J. Am. Chem. Soc.*, 129 (2007)2559.
[23] Y. Shimazaki, T. D. P. Stack, and T. Storr, *Inorg. Chem.*, 48 (2009)8383.
[24] M. Orio, O. Jarjayes, H. Kanso, C. Philouze, F. Neese, and F. Thomas, *Angewandte Chemie - International Edition*, 49 (2010)4989.
[25] T. Storr, E. C. Wasinger, R. C. Pratt, and T. D. P. Stack, Angewandte *Chemie - International Edition*, 46 (2007)5198.

[26] A. Hamnett, J. Abel, J. Eameaim, P. Christensen, A. Timonov, and S. Vasilyeva, PCCP, 1 (1999)5147.
[27] B. Bag, N. Mondal, G. Rosair, and S. Mitra, *Chem. Commun.,* (2000)1729.
[28] Y. Shimazaki, F. Tani, K. Fukui, Y. Naruta, and O. Yamauchi, *J. Am. Chem. Soc.*, 125 (2003)10512.
[29] O. Rotthaus, F. Thomas, O. Jarjayes, C. Philouze, E. Saint-Aman, and J. L. Pierre, *Chem. Eur. J.*, 12 (2006)6953.
[30] J. Tedim, S. Patricio, J. Fonseca, A. L. Magalhaes, C. Moura, A. R. Hillman, and C. Freire, *Synth. Met.,* 161 (2011)680.
[31] O. V. Levin, M. P. Karushev, A. M. Timonov, E. V. Alekseeva, S. H. Zhang, and V. V. Malev, *Electrochim. Acta,* 109 (2013)153.
[32] V. V. Malev, O. V. Levin, and A. M. Timonov, *Electrochim. Acta*, 108 (2013)313.
[33] E. A. Dmitrieva, S. A. Logvinov, V. V. Kurdakova, V. V. Kondrat'ev, V. V. Malev, and A. M. Timonov, *Russ. J. Electrochem.*, 41 (2005)381.
[34] M. F. Mathias and O. Haas, *J. Phys. Chem.*, 96 (1992)3174.
[35] S. A. Krasikova, M. A. Besedina, M. P. Karushev, E. A. Dmitrieva, and A. M. Timonov, *Russ. J. Electrochem.*, 46 (2010)218.
[36] G. A. Shagisultanova, M. E. Ivanova, I. E. Popeko, and A. M. Timanov, *Zh. Neorg. Khim.,* 36 (1991)3096.
[37] A. Hamnett, J. Abel, J. Eameaim, P. Christensen, A. Timonov, and S. Vasilyeva, *Phys. Chem. Chem. Phys.*, 1 (1999)5147.

7. Possible Developments of the Polaron Conductance Representations

From the above discussion of a series of experimental studies performed with conducting polymers (CPs) it is clear that the results of such works are only in a qualitative agreement with the existing theoretical representations on charge transport within electroactive polymers. Not repeating the previous arguments in favor of such affirmation, we would like to restrict ourselves only to the indication that, among problems to be solved in connection with an adequate description of CPs, at least two ones seem to be of the most importance.

The first problem originates from the established presence of some conjugation between the charge and solvent fluxes into polymer films. In our opinion, the problem can be solved in the way applied usually in the non-equilibrium thermodynamics, according to which linear combinations of gradients of electrochemical potentials of transferring species are considered as moving forces for the fluxes of such particles [1]. However, something new should be introduced into this approach in order to describe the above indicated sign conversion in the solvent fluxes at changing the oxidation degree of PEDOT and poly-Ni(Schiff) films (see Sections 2, 4). It seems that the desired solution might be achieved only if the appearance of mechanical strains in a polymer film with variations in its oxidation is accounted for at a proper consideration. This is not a simple task, but might issue in a practical importance connected with the artificial muscle problem.

The second problem is less obvious than the first one and, as mentioned earlier, springs from some doubts in the accepted approach to treating CPs as practically exact analogs of redox polymers (see Section 2). To explain this, we remind the reader that, in scope of the widely used homogeneous model of electroactive polymer films, the corresponding flux equations are as follows

$$J_1 = - D_1\{\partial C_1/\partial z - C_1(1-\theta)\,\partial[\Phi - a\theta]/\partial z\}$$
$$J_2 = - D_2\{\partial C_2/\partial z + z_2 C_2 \partial\Phi/\partial z\} \qquad (7.1)$$

Here, the used notations are determined by the following definitions: $C_1(z,t)$ and $C_2(z,t)$ are the concentrations of reduced film fragments and counter-ions, respectively; $\theta = C_1/N$ is the occupancy of the film fragments (sites) by their reduced form; N is the total concentration of film sites that is not dependent on distance z from the substrate/film interface; $\Phi(z,t) = F\varphi(z,t)/RT$, the dimensionless electric potential ($\varphi(z,t)$, its value in electrical units (here V)); a, the so-called attraction constant (in RT units) characterizing short-range interactions between film fragments; D_k, the diffusion coefficient of the k-th species (k =1, 2); J_k, their flux; z_2, the charge of counter-ions; t, the time variable [2-4].

We also remind that, according to the above definitions, the occupancy of film fragments with their oxidized form is equal to $(1-\theta)$, which corresponds to the fact that any film site might be in either the oxidized or reduced state. In other words, the so-called "hopping" mechanism of charge transfer takes place within the film interior [2-4], *i.e.* the electron exchange reaction between two neighboring film fragments of different oxidation states (Red and Ox) is as follows

$$\text{Red} - e \leftrightarrow \text{Ox} \qquad (7.2)$$

The same set of flux equations (7.1) is assumed be valid for the case of polaron conducting films [2-4]. In reality, this assumption means an introduction of the additional one, namely the supposition that all the reduced fragments participating in the polaron formation reaction

$$m\,\text{Red} - e \leftrightarrow P \tag{7.3}$$

should be considered as a separate site (or a separate quasi-particle) in the film lattice (P symbolizes a polaron quasi-particle, m is the number of film fragments included into the polaron structure). As pointed out in review [5], "imperfection of such consideration becomes obvious if any two polaron particles of a polymer chain are divided with the reduced ones, the number of which is smaller than that included into the polaron structure; in scope of the assumption in question, the polaron transfer is forbidden under these conditions". Not detailing this statement (other details see in [5]), we indicate only that, in any case, the above additional assumption seems questionable and, maybe, it is an excessive one. In this context, the justified question might arise whether some quantitative changes in the polaron conductance description should appear in result of refusing from such assumption? As will be shown below, it actually happens.

7.1. Quasi-Equilibrium Cyclic Curves of CP Films Containing M-Site Polarons

Let us consider a modified electrode covered with a conducting polymer film and immersed into a symmetrical (1 – 1) electrolyte solution of concentration C_S in its bulk. Assuming any reduced (further on, neutral) fragment of polymer chains to be a separate particle (but not their complex consisting of m fragments), one can apply the Gibbs principle of equilibrium [6] to reaction (7.3):

$$m\mu_{Red}(f) - \tilde{\mu}_e(es) = \tilde{\mu}_P(f), \tag{7.4}$$

where $\mu_{Red}(f)$ is the chemical potential of neutral (reduced) fragments in the film bulk, while $\tilde{\mu}_P(f)$ and $\tilde{\mu}_e(es)$ are the electrochemical potentials of polarons (in the film bulk) and electrons (in the electrode substrate), respectively. For the sake of simplicity of subsequent derivations, one can consider the film as an ideal mixture of polarons and reduced fragments, *i.e.* neglect specific interactions between the film particles. This leads to the following equilibrium relationship:

$$\theta_0/(1-\theta_0)^m = \exp\{[m\mu^0_{Red}(f) - \mu^0_P(f) - \mu^0_e(es)]/RT\}\exp(E-\Phi_0) == k_e\times$$

$$\times\exp(E-\Phi_0) \tag{7.5}$$

Here, $\mu^0_P(f)$, $\mu^0_e(es)$, $\mu^0_{Red}(f)$ are the standard chemical potentials of the corresponding species; E and Φ_0 are the dimensionless electric potentials (in RT/F units) of the substrate and the film bulk (with respect to the solution bulk potential); θ_0 is the equilibrium occupancy of the film lattice with polaron fragments, so that factor $(1-\theta_0)$ gives the probability of finding of a sole reduced fragment in the film interior; $k_e = \exp\{[m\mu^0_{Red}(f) - \mu^0_P(f) - \mu^0_e(es)]/RT\}$ is the partition coefficient of electrons between the film and the electrode substrate.

According to the initial assumption, any polaron quasi-particle includes m fragments and, therefore, their equilibrium concentration in the film interior, C_1^0, is equal to $N\theta_0/m$, which, after accounting for Eqn. (7.5), gives

$$C_1^0 = N\theta_0/m = (Nk_e/m)\,(1-\theta_0)^m \exp(E-\Phi_0) \tag{7.6}$$

The latter reduces to the well-known result for concentration $N\theta_0^0$of positively charged (oxidized) film fragments [2-4]:

$$C_1^0 = N\theta_0^0 = Nk_e\,(1-\theta_0^0)\exp(E-\Phi_0^0) \text{ at } m=1, \tag{7.6'}$$

if one puts number m be equal to unity (upper index "0" is used in Eqn. (7.6') to emphasize possible discrepancies of the quantities marked with the index from those included into Eqn. (7.6)). Taking into account the indicated difference in Eqns. (7.6, 7.6'), one can actually expect the appearance of some new results (as compared to the known ones) in scope of the proposed consideration. To continue it, we indicate that the equilibrium counter-ions concentration in the film bulk, C_2^0, is given by the equation [2-4]:

$$C_2^0 = k_m C_S \exp(z_2\Phi_0) \tag{7.7}$$

independently of the fact what number m is assigned to polaron quasi-particles (k_m, the partition coefficient of counter-ions; C_S, their concentration in the bathing electrolyte bulk). As mentioned earlier, the electroneutrality condition:

$$C_1^0 + z_2 C_2^0 = 0 \tag{7.8}$$

should be satisfied in the most part of the film interior, so that, after substitutions of Eqns. (7.6, 7.7), this condition takes a view

$$(Nk_e/m)\cdot(1-\theta_0)^m\cdot\exp(E-\Phi_0) + z_2\,k_m\,C_S\exp(z_2\Phi_0) = 0 \tag{7.9}$$

The θ_0-value included into this relationship must be found from Eqn. (7.6) in order to obtain finally the desired dependence of film bulk potential Φ_0 on electrode one E. It appears that this task has an analytical solution if a polaron quasi-particles contains only two monomer (or repeat) units (*i.e.* m = 2) and charge $z_2 = -1$. In this case, Eqn (7.6) is a quadric algebraic one, so that occupancy θ_0 might accept two values given below

$$\theta_0 = 1+\exp(\Phi_0 - E)/2k_e \pm \exp(\Phi_0 - E)/2k_e\,[1+4k_e\exp(E-\Phi_0)]^{1/2}, \tag{7.10}$$

Substituting this expression into the electroneutrality condition (Eqn. (7.9) rewritten for the considered case of m =2 and $z_2 = -1$), one obtains the following equation

$$N/2 + N\exp(\Phi_0 - E)/4k_e \pm [\exp(\Phi_0 - E)/4k_e]\cdot[1+4k_e\exp(E-\Phi_0)]^{1/2} = \\ =k_mC_s\exp(\Phi_0) \tag{7.11}$$

If one becomes free from the radical of this equation, the latter reduces to $\exp(2\Phi_0) - (N/k_mC_s)[1-(N/2k_ek_mC_s)\exp(-E)]^{-1}\exp(\Phi_0) + ((N/k_mC_s))^2\cdot[1-(N/2k_ek_mC_s)\exp(-E)]^{-1} = 0$ and, hence, film bulk potential Φ_0 might also accept two values determined by the expression:

$$e^{\Phi 0} = (N/2k_mC_s)[1-(N/2k_ek_mC_s)e^{-E}]^{-1}\{1\pm(N/2k_ek_mC_s)^{1/2}e^{-E/2}\} \tag{7.12}$$

This allows one to write counter-ion concentration C_2^0 and then film charge $Q = FLC_2^0$ in explicit forms as follows

$$C_2^0 = N/2[1\pm\exp(U/2)] \tag{7.13}$$

and $Q(U) = FLN/2[1 \pm \exp(U/2)]$ (7.14)

where new independent variable U included into these equations is defined by the equality: $U = u_0 - E = \ln(N/2k_ek_mC_s) - E$.

In scope of cyclic voltammetry, potential E is a linear function of time t and, therefore, current density $I(t) = dQ/dt$ appears be equal to

$$I(t) = I[U(t)] = \pm (\nu F^2 LN/4RT)\cdot\exp(U/2)/[1 \pm \exp(U/2)]^{2,} \quad (7.15)$$

where ν, the rate of sweeping the electrode potential. The routine analysis of the above expression permits one to establish, first, that, among two possible signs in Eqns. (7.13-15), only positive one has a real (physical) meaning. Thus, Eqn. (7.15) should be rewritten in the form:

$$I(t) = I[U(t)] = \pm (\nu F^2 LN/4RT)\cdot\exp(U/2)/[\exp(U/2) + 1]^{2,} \quad (7.15')$$

where sign "+" conforms to the anodic direction of sweeping the electrode potential, while the negative sign is assigned to the cathodic direction. Second, current dependence I(U) possesses two symmetrical extrema appearing at $U_{max} = 0$ (i.e. $E_{max} = \ln(N/2k_ek_mC_s)$) and corresponding to different directions of sweeping the electrode potential. The absolute value of such peak currents, $|I_p|$, is given by the equation

$$|I_p| = \nu F^2 LN/16RT \text{ at } m = 2 \quad (7.16)$$

It should be indicated here that Eqn. (7.15') in its type is close to the Laviron equation of cyclic voltammetry with electrodes covered by compact monolayer films of an adsorbate capable to reduction/oxidation [7, 8] (see also [9]). The main difference between such equations is in the fact that variable U (*i.e.*, in essence, electrode potential E) times factor ½ is included into Eqn. (15′), while the Laviron one contains the analogous variable without such factor. This is the reason why semi-width h of the cyclic curve described by Eqn. (7.15') appears to be exactly two times higher (namely, h = 181 mV at room temperature) than that follows from the Laviron equation (h = 90, 6 mV).

What is important, the obtained quasi-equilibrium cyclic curve (that is Eqn. (7.15')) differs not only from the Laviron one, but also the curve that follows from the analogous consideration in case of m =1 (and $z_2 = -1$), *i.e.*, in essence, the case considered in the existing approach to treating the polaron conductance of polymer films. The corresponding curve was calculated in [10] (case of *p*-doped polymers) in the way similar to that used above. Here, it is sufficient to point out that the peak current density values of such curve are given by the equation:

$$|I_p| \approx 0.172 \nu F^2 LN/RT \text{ at } m = 1, \tag{7.17}$$

while its semi-width h appears be equal to 131 mV at room temperature [10]. A subsequent comparison between these results (at m =1) and the preceding ones (at m =2) must obviously be performed at the same values of total charge Q consumed at charging both model systems. To satisfy such requirement, one should decrease the peak current density of Eqn. (7.17) by two times exactly:

$$|I_p| \approx 0.086 \nu F^2 LN/RT \text{ at } m=1, \tag{7.17'}$$

in order to equalize maximum polaron occupancies ($\lim_{E \to \infty} \theta_0^0$ and $\lim_{E \to \infty} \theta_0$) in the both cases considered. As charge Q is the product of the peak current and the semi-width of the corresponding cyclic curve, the ratio of charges Q(m=1)/Q(m=2) actually coincides with unity at the conditions indicated:

$$Q(m=1)/Q(m=2) \approx 0.086 \cdot 131 \cdot 16/181 \approx 1 \tag{7.18}$$

The established differences in the peak currents and the semi-widths for the compared cases are obviously reserved at the equalizing procedure performed. As the above parameters of the compared cyclic curves significantly differ from each other, it is possible to conclude that the proposed and existing approaches to treating the polaron conductance lead to the essentially different quantitative results, at least in the case of two-site polarons. Although the above comparison has been carried out in the partial case of m = 2, one can reasonably assume that differences in the quantitative inferences following from the compared approaches must be the greater the greater number m of the repeat units included into the polaron structure.

Thus, the expected appearance of new results actually takes place in scope of the proposed approach even if the level of equilibrium and quasi-equilibrium relationships is considered.

7.2. Probable Directions of Theoretical Studies of CPs

The approach shortly outlined above needs some further development and its inferences should be verified with the corresponding experimental studies. In this context, it seems necessary to extend the proposed description on the case of non-equilibrium conditions, for example those, which take place at impedance measurements. Indeed, the cyclic voltammetry method gives restricted information on the electrode processes occurring in CPs and, therefore, it might be applied only at their primary characterization (see Chapter 2). This is the reason, why the results obtained in the preceding paragraph can be considered only as some premise to more detailed considerations. Here, it must be pointed out that we have performed the necessary analysis of some partial models of polaron-containing films in non-equilibrium conditions. Not detailing the obtained results (they will be published elsewhere), we indicate that the derived polaron flux equation differs from that used (in essence, the first equation of set (7.1)) in the existing theory of modified electrodes. As it must be, the same expression for the polaron concentration (see Eqn. (7.6)) follows from the obtained flux equation at equilibrium conditions, which is important, since it shows self-consistency of the applied approach. These circumstances mean the inevitable appearance of new issues at the level of non-stationary processes in polaron-conducting polymer films.

As concerns experimental verifications of the outlined approach, at the given stage of its development they seem to be premature, since a series of real features of CPs should preliminarily be accounted for. Among such features, it is necessary to analyze firstly possible effects of short-range-interactions within the objects in question. On the level of equilibrium and quasi-equilibrium results, the necessary analysis can be performed along the above-stated way with using the traditional representation of the electrochemical potentials of transferred particles in the Bragg-Williams (or molecular field) approximation [11]. This task can easily be solved, but requires numerical calculations (see, for example [10, 12]). We would not like to detail further a possible list of the desired developments, but indicate only the following problem to be solved. It originates from some uncertainty in the number of repeat polymer units included into the polaron structure; as mentioned above, this number varies from three to six depending on the polymer nature and, probably, the solvent used. One cannot exclude that some population of polarons with different numbers of containing units might exist in real polymer films due to their local inhomogeneity. If so, the experimentally determined numbers of polaron fragments should be treated as mean values averaged over the existing polaron population. This means that the adequate description of such systems must account for possible changes in polaron sizes, to say nothing about polaron/bipolaron transformations in result of the so-called disproportion reactions [13]. In our opinion, the resolution of this problem will help to elucidate, at least the concrete reasons responsible for the existence of several current peaks in cyclic curves of electrodes covered with the Schiff base polymer complexes (see Section 6).

It should also be emphasized here that, from the general point of view, a quantitative treatment of the polymer film conductance should be performed *ab ovo*, *i.e.* in terms of the quantum mechanics, since some changes in the electronic coupling between separate fragments of polymer chains underlie charge transport within these objects. This is the reason why phenomena related to charge transfer in single molecules and microscopic wires (or their nets) formed such species have been the subjects of numerous studies during the last years (for example, see [14, 15]). One can therefore expect that the further development of such approaches will also assist in elucidation of the problems discussed in this review.

References to Section 7

[1] S. R. de Groot and P. Mazur, Non-Equilibrium Thermodynamics, *Mir*, Moscow, 1964 (Russian translation).

[2] N.A. Skotheim, R.J. Elsenbanmer, J.R. Reynolds, *Handbook of Conducting Polymers*, Marcel Dekker, NY, 1998.

[3] G. Inzelt, *Conducting Polymers. A New Era in Electrochemistry*, Springer-Verlag, Berlin, Heidelberg, 2008.

[4] V.V. Malev, V.V. Kondratiev, A.M. Timonov, *Polymer-Modified Electrodes,* Nestor-history, St. Petersburg, 2012 (in Russian)

[5] V. V. Malev, O.V. Levin, V.V. Kondratiev, *Electrochimica Acta* 122 (2014) 234.

[6] J. W. Gibbs, *Thermodynamics. Statistical Mechanics*, Nauka, Moscow, 1982 (Russian translation).

[7] E. Laviron, *J. Electroanalyt. Chem*. (Interfacial Electrochem.) 52 (1974) 355.

[8] E. Laviron, in *Electroanalytical Chemistry*, V. 12, A.J. Bard (Ed.), Marcel Dekker, NY, 1982, P. 53.

[9] A. J. Bard, L. R. Faulkner, *Electrochemical Methods. Fundamentals and Applications*, John Wiley&Sons Inc., NY, 1980.
[10] V. V. Malev, O.V. Levin, *Electrochimica Acta* 80 (2012) 426.
[11] E. A. Moelwyn-Hughes, Physical Chemistry, V.2, Ed. Ya. I. Gerasimov, *Foreign literature,* Moscow, 1962, (Russian translation)
[12] V. V. Malev, O. V. Levin, M. A. Vorotyntsev, *Russian Journal of Electrochemistry* 43 (2007) 1016.
[13] V. V. Malev, O. V. Levin, A. M. Timonov, *Electrochimica Acta* 108 (2013) 313
[14] A. A. Kocherzhenko, Ferdinand C. Grozema and Laurens D. A. Siebbeles, *Phys. Chem. Chem. Phys.*, 13 (2011) 2096.
[15] I. I. Fishchuk, A. Kadashchuk, H. Bassler, S. Nespurek, *Physical Review* B, 67 (2003)224303

CONCLUSION

Although the electrodes modified with the films of polythiophenes and polymer nickel complexes with Schiff base ligands possess different properties, these electrode systems reveal a certain resemblance in their electrochemical behavior. This firstly concerns the established polaron-ion character of their conductance. The second common feature is in the fact of conjugation between the fluxes of charge-compensating ions and solvent molecules of the electrolytes used. Moreover, both electrode systems demonstrate some expansion of modifying films with an increase in their oxidation degree. Finally, the experimental results obtained for the both systems are only in a qualitative agreement with the accepted theory of charge transfer within electroactive polymer films. In our opinion, such incomplete agreement results from the absence of principal differences in the existing representations on the conductance of the redox and polaron-containing polymer films. However, some discrepancy appears if the reduced repeat units of polymer chains are assumed to be separate particles instead of their complexes consisting of three-six units and considering as separate quasi-particles in scope of the existing theory. Though such discrepancy, at first sight, seems immaterial, it leads to quantitative changes in the inferences following from the proposed and accepted approaches to treating the polaron conductance of polymer films. Additional developments of the proposed approach should precede its experimental verifications.

One more conclusion should be done in connection with the performed studies of electrode processes at PEDOT films containing the noble metal inclusions. The obtained experimental results are in agreement with the previously developed treatment of electrochemical processes in composite electrodes and show the possibility of using such systems at electroanalytical determinations of different substances.

ACKNOWLEDGMENT

Financial support of the Russian Foundation for Basic Research (grants ## 13-03-00843-a, 12-03-00566, 13-03-00984 and 14-29-04057) and the St. Petersburg State University (grant # 12.38.77.2012) is gratefully acknowledged. SEM and TEM studies were performed using the facilities provided by the Interdisciplinary Resource Center "Nanotechnology" of St. Petersburg State University.

In: Advances in Conducting Polymers Research
Editor: Laura Michaelson
ISBN: 978-1-63463-258-4

Chapter 5

CONDUCTING POLYMER-FUNCTIONALIZED CARBON NANOTUBES HYBRID NANOSTRUCTURES BASED BIOANALYTICAL SENSORS

***Sushmee Badhulika*[1] *and Ashok Mulchandani*[2]**
[1]Department of Electrical Engineering,
Indian Institute of Technology, Hyderabad, India
[2]Department of Chemical and Environmental Engineering,
University of California, Riverside, US

ABSTRACT

Sensors form an integral part of our everyday lives in a wide range of disciplines ranging from detection of environmental toxins, quality control in food and water to healthcare and general safety. Nanomaterials such as carbon nanotubes (CNTs) owing to their small size, high electrical and thermal conductivity, high specific area and superior electronic properties are strong candidates for analyte detection and are thus being increasingly incorporated in sensor architecture. The electrically conducting polymers (CPs) are known to possess numerous features in terms of stability and ease of processing. Their high chemical sensitivity, room temperature operation and tunable charge transport properties has made them ideal for use as transducing elements in chemical sensors. Utilizing the property of surface modification of CNTs, CPs-CNT hybrid structures have been developed by electropolymerization. These hybrid structures exhibit the synergistic benefits of both the materials and allow rapid electron transfer for the fabrication of efficient sensors. This chapter focuses on the synthesis, characterization and applications of conducting polymer-CNTs hybrid nano bio/chemical sensors in various modes of sensor configurations towards sensing gases; volatile organic compounds (VOCs) and biomolecules whose detection and analysis plays a crucial role in environmental pollution control, medical diagnostics and food safety.

1. INTRODUCTION

With increasing awareness of the public vulnerability to chemical and biological agents, there is a heightened need for detection techniques that show both high sensitivity and selectivity. Such techniques would also find wide spread use in medical diagnostics, biomedical research applications, forensics, and homeland security. Hence chemical sensors form an integral part of research initiatives in a wide range of disciplines ranging from detection and control of environmental toxins, quality control in food and water and general healthcare and safety. Chemical sensors come in a broad category depending upon their construct, properties and applications. But overall they can be defined as devices that transforms chemical information into an analytically useful signal. The main requirements of a good sensor are high sensitivity, fast response, low cost, high volume production, and high reliability that can be operated on site and in situ. With time, the need for miniaturization of these sensors has been realized to complement the ever changing technological demands. This led to the emergence of nanotechnology with a purpose aimed at a strategic shift in sensor technology towards more sensitive recognition layers, increasingly complex architectures and reduced size and more so because of the fact that silicon-based semiconducting metal oxide technologies will reach its limit in the near future.

In the past few decades, several approaches have been developed and demonstrated to fabricate simple, robust solid state sensors whose operation is based on the energy conversion of binding of an analyte at the active surface of the sensor to a measurable signal (resistance, capacitance, temperature etc.). However, with the advent of nanotechnology, the current goal has been replaced to the fabrication and successful demonstration of nanosensor devices, which when integrated with modern electronic fabrication technologies delivers faster, better, low cost and low power operations.

At the nanometer scale, the wave-like properties of charge carriers inside matter and atomic interactions are greatly influenced by the size of the material. As a result, changes in the size dependent properties (e.g. melting points, magnetic, optical, and electronic properties) may be observed even without any significant compositional change. Furthermore, their high surface-to-volume ratio results in a tremendous improvement in chemical properties thereby making these nanostructures highly sensitive to any surface phenomena. Besides, the confinement-induced quantization effect could also occur when the size of materials becomes comparable to the de Broglie wavelength of electrons inside the material [1]. By creating nanostructures, therefore, it is possible to control the fundamental properties of materials through the surface/size effect. This in turn allows us to develop new materials and advanced nano devices of desirable properties and functions.

Rapid progress in the synthesis and fundamental understanding of surface phenomena of nanomaterials has enabled their incorporation into sensor architectures. Nanomaterials are strong candidates for analyte detection, because their reduced dimensions creates an increase in environmental sensitivity. The reduced dimensionality also creates structures with exceptionally high surface area, and some materials, such as certain types of carbon nanotubes (CNTs) which are composed almost entirely of surface atoms [2]. These two consequences of reduced size result in a class of materials that has the potential for unsurpassed sensitivity towards changes in its chemical environment thus making them promising candidates for sensing applications. To address specific requirements demanded for particular sensing applications it is therefore required that nanomaterials be combined with one or more materials such as metals, metal oxides and conducting polymers that impart additional beneficial properties to them. Techniques for such chemical modification of nanostructures have been proposed and successfully implemented.

1.1. Significance of One Dimensional Structures over Thin Film Structures

In a conductometric sensor, when a charged/ionic analyte binds to the sensing element of a sensor it undergoes an addition/depletion in the number of charge carriers thereby leading to the change in the resistance/conductance of the sensing material. In a conventional thin film based sensor only a fraction of the total cross-sectional area is affected by such modulation in the number of charge carriers. As a result, the current shunts across such regions of modified conductance and leads to depletion in the sensing signal. On the other hand, owing to their small sizes, one-dimensional nanostructures like nanowires/nanotubes show modification in the electrical properties across its entire cross-section and allow surface chemistry to play a role as compared to in bulk state (Figure 1.1.1). This ballistic conduction of current through the nanostructures results in an enhanced signal and thus makes them more sensitive than the thin film sensors. Furthermore, the higher sensitivity of such 1-D nanostructures can be attributed to the following reasons:

(a) A large surface area-to-volume ratio which means that a significant fraction of the atoms (or molecules) in such systems are surface atoms that can participate in surface reactions.
(b) The Debye length λ_D (a measure of the field penetration into the bulk) for most nanowires and nanotubes is comparable to their radius, which causes their electronic properties to be strongly influenced by processes at their surface, thus resulting in better sensitivity.

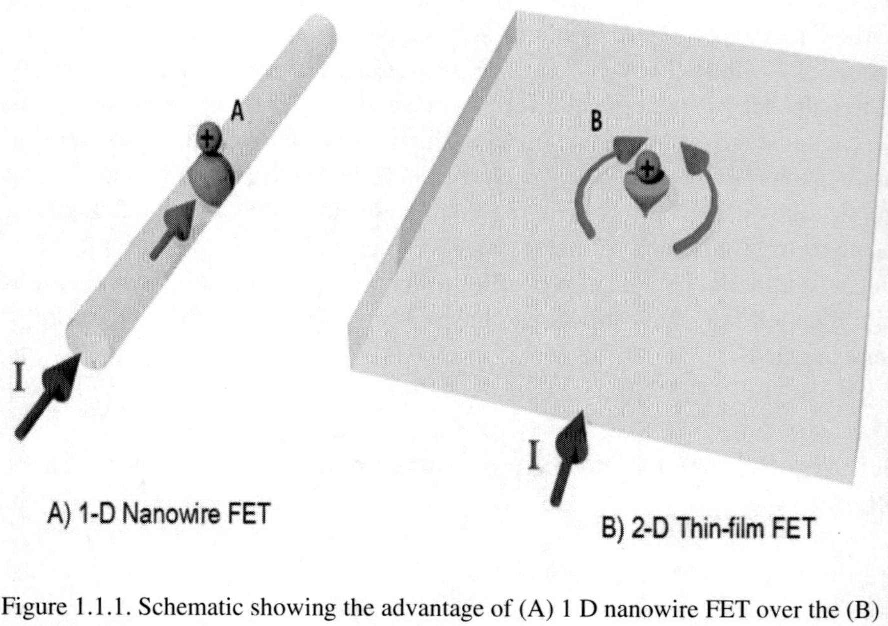

Figure 1.1.1. Schematic showing the advantage of (A) 1 D nanowire FET over the (B) 2-D thin film FET.

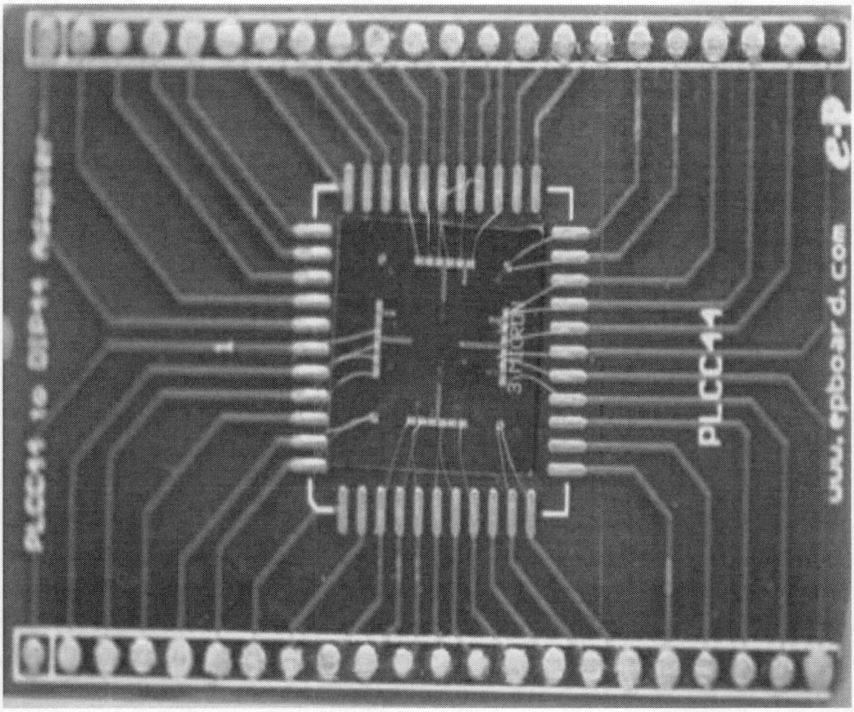

Figure 2.1. 16 sensors array on chip carrier.

Finally, nanowires and nanotubes can be easily configured as field-effect transistors (FETs) and can be potentially integrated with electronic circuits, conventional devices and device fabrication techniques.

1.2. Single-Walled Carbon Nanotube as Sensing Element

A single-walled carbon nanotube (SWNT) is a 1-D hexagonal network of carbon atoms of approximately 1 nm diameter and 1 to 100 microns length. A SWNT can be metallic, semimetallic or semiconducting depending on its chirality. SWNT displays excellent chemical stability, good mechanical strength and a range of electrical conductivity [2] that makes then ideal building blocks for electronic applications. Its properties of small size, high strength, high electrical and thermal conductivity and high specific area have been successfully explored and used in applications related to scanning probes, actuators and nanoelectronic devices [3-6]. When compared with other conductance sensitive materials like metal oxide semiconductors and conducting polymers, SWNT offers advantages in terms of higher sensitivity, smaller sizes that facilitates fabrication of miniaturized sensors and massive nanosensor arrays, lower power consumption for wireless applications. Compared to silicon piezoresistors [7], SWNTs have proven to be less sensitive to variations in temperature. This makes them perfect candidates for chemical sensors and also as highly sophisticated implantable sensors. The remarkable sensitivity of SWNT conductivity to the surface adsorbates permits the use of SWNT as highly sensitive nanoscale sensors. The sensing behavior arises from the fact that changes in their local chemical environment has a direct impact on their electronic structure and hence provides unambiguous path for detecting interacting molecules that initiate the change. SWNTs have been extensively used in the field of bio and gas sensing.

1.2.1. Carbon nanotubes in gas and biosensing

Biosensors are the class of chemical sensors in which the recognition system utilizes a biochemical mechanism or in other words incorporates a biological sensing element connected to a transducer. They could be used to detect a wide variety of analytes ranging from levels of acidity, alcohol, minerals, starch, toxins and vitamins in dairy, fish, meat and beverages. Based on the analyte to be detected, typical chemistries are used for necessary detection and quantification. The transduction element of a biosensor performs the function of converting a specific biological reaction (binding or catalytic) into a response that can be processed into a useable signal. This element must also be suitable for the immobilization of the biological component at, or close to its surface. SWNTs in the field of biosensing enable the use of new signal transduction technologies. Their high surface area makes it possible to enhance the amount of immobilized biomolecules in sensor construction thus improving the limit of detection of a sensor. The combination of excellent conductivity, good electrochemical properties, and nanometer dimensions allows them to being plugged directly into individual redox enzymes for better transduction in electrochemical enzyme biosensors [8-10]. The direct electron transfer ability of carbon nanotubes has also been exploited in research findings [11] wherein use of SWNT has made possible a direct electron transfer with the redox active centers of adsorbed oxidoreductase enzymes. SWNTs modified with bio recognition molecules, such as antibodies, aptamers or DNA, have been successfully used to detect various targets, including proteins, bacteria, yeast, DNA/RNA, mammalian cancer cells [12-14] and small and weakly charged molecules like glucose [15].

Similarly, SWNTs have been found to be ideal building blocks for making gas sensors. The SWNTs consists solely of surface atoms such that every single carbon atom is in direct contact with the environment which makes them highly sensitive to any surface adsorption event. The earliest reports of gas sensors made from semiconducting SWNTs showed good sensitivities toward NH_3 and NO_2 at room temperature. [16]. The resistance response of SWNTs to gas adsorption has been detected by employing them in field emit transistors (FETs) mode of configuration [16-18]. If two metal contacts are connected to each end of an individual semiconducting SWNT, the metal/S-SWNT/metal device exhibits p-type transistor characteristics. Someya et al. characterized the influence of alcohols adsorption on the resistance of SWNTs using the same transistor structure [17]. The responses were found to be reversible and reproducible over many cycles of vapor exposure. Since then there have been numerous reports to investigate and understand characteristics such as reversibility, reproducibility, sensitivity, and selectivity to various gaseous analytes.

However, gases which have a low absorption capacity on the CNTs surface were difficult to detect. This less than ideal sensitivity and lack of selectivity to detect many types of analytes such as volatile organic compounds (VOCs) and gases due to weak binding energy of the nanotubes necessitated the modification of SWNTs with certain compounds like metals, oxides and polymers etc. When functionalized with materials like conducting polymers [19, 20], metals [21-24] and metal oxides [25], SWNTs yielded superior sensing behavior. Similarly, for the detection of biomolecules that are weakly adsorbed on SWNTs surface, the change in resistance is often too small, and thus, their non-covalent [26] or covalent [27, 28] modification is needed.

1.3. Conducting Polymers and Their Application in Gas Sensing

Conducting polymers (CPs) are an attractive group of materials which exhibit electronic, magnetic and optical properties of metals and semiconductors while retaining the attractive mechanical properties and processing advantages of polymers [29, 30]. They possess the natural length of polymer chains and their morphologies in the bulk lies precisely at the nanometre length scale. [31].

The large number of possible conformations available to a polymer chain on a surface and/or in the bulk, indicates considerable room for creating polymeric materials of new properties and functions even without any change in their chemical composition. CPs consist of repetitive sequences of monomer units that contain sp^2 hybridized backbones. Incorporation of dopants onto them results in change in their conductivity levels depending upon the dopant type and its concentration. Their conductivity can be varied as high as fifteen orders of magnitude by varying the dopant level thus making them excellent flexible building materials for wide range of electronic applications [30]. This prominent feature of CPs along with high chemical sensitivity, room temperature operation and tunable charge transport properties has made them ideal for use as transducing materials in chemical sensors [32, 33].

CPs have been put to use in form of thin and thick film sensors for detection of vapours and gas analytes [31, 34]. Nano sensors based on CPs exhibit the property of bulk penetration of analytes thus making them highly sensitive. The earliest reports of conducting polymers and their utility in the field of gas sensing came in the form of thick/thin films of polypyrrole (PPy), polyaniline (PANI), polythiophene (PT), and their derivatives with a clear understanding of the of interaction of the gases with the CPs structures. While in some cases swelling of the polymer film [35] occurred due to interaction with gas molecules thereby resulting in an increase in the average hopping distance between chains that led to an increase in the resistance of the sensor device, in other cases it resulted in a decrease in resistance caused due to change alignment [36, 37]. The dopant dependent conductivity together with the porosity of the CP enabled electrochemical activity thus allowing redox control over conductivity. This facilitated ease of processing of CP using electrochemical techniques.

However CPs do have certain disadvantages which limits their application in long term use.

When used as films, they lack the 1-D nanostructures' advantages. To overcome this drawback, one dimensional CP nanowires were synthesized. Some of the first attempts involved single polymeric polyaniline nanowire based gas sensor which showed a rapid and reversible resistance change upon exposure to NH_3 gas at concentrations as low as 0.5 ppm [38]. Template directed PPy nanowire; both drop casted and dielectrophoretically positioned between prefabricated microelectrodes were also synthesized and demonstrated for the detection of ammonia [34]. Further enhancement in the structure of the polymer nanowires was achieved by synthesis of Au/PEDOT/Au nanowire [39] which were then used for the detection of volatile organic compounds. These gas sensors showed low sample to sample variation and proved to exhibit excellent reproducibility. However, their diameters were found to be much larger than those of SWNTs. Another disadvantage of CPs was that over time they oxidized when exposed to oxygen.

Thus it was essential to broaden the scope of CNTs and CPs by using both materials in concert to create a hybrid which would have a higher sensitivity than either does on its own. In addition, by changing the dopant in CPs, different selectivities could be introduced to the resulting material. Taking into considerations the demands of sensor industries for small size and highly efficient sensors which show promise of better sensitivity and selectivity, hybrid sensors were fabricated that incorporated CNTs as the main building block functionalized with CPs as the sensing element. The surface functionalization concerning conducting polymers and carbon nanotubes is a very broad field that encompasses concepts of material synthesis, structural characterization and determination, surface/interface design and modification. Polymer–nanotube hybrid composites constitute a new class of hybrid nanomaterials, which show characteristic properties of both constituent components with potential synergetic effects. It has been long recognized that surface phenomena play a critical role in the functioning of a broad range of materials and devices. With so many advanced surface functionalization techniques and device fabrication methods already existing and more to be developed, the examples discussed in this section, mainly deal with electropolymerization of conducting polymer on the surface of aligned CNT networks (that summarize our recent research activities along this line) and SWNTs-CPs based composites for gas sensing.

1.4. Scope and Overview

This chapter deals with the synthesis, characterization and operation of CNTs and CPs based hybrid nanosensors towards detecting gases; volatile organic compounds (VOCs) and biomolecules. The combination of CNT and CPs offers a very promising and attractive route for synthesis of novel, hybrid compounds that can be used in a variety of applications ranging from gas sensors, biosensors, actuators, and molecular scale electronic devices. This chapter mostly deals with the development of such sensors and investigates their utility in bioanalytical applications with focus mainly on our own work involved in this direction. So, there is no doubt that the coverage of comprehensive literature in this chapter will not exhaust all significant research findings reported elsewhere.

2. Sensor Design and Fabrication

Microfabricated gold electrodes can be fabricated on highly doped p-type silicon substrate using the cleanroom tools and techniques. The procedure includes deposition of a 300 nm SiO_2 thick film on a (100) oriented highly doped p-type Si substrate by thermal CVD deposition to insulate the substrate. It is then followed by defining the drain and source electrode areas by photolithography using a positive photoresist such as 5214. For the purpose of using it in our research, the thickness of the chromium adhesion layer was maintained at 20 nm and that of gold above it was 180 nm. The width and gap of the electrodes were fixed at 200 μm and 3 μm, respectively. The chip configuration was such that it contained 16 addressable electrode pairs as shown in Figure 2.1.

Several studies have been done to develop techniques through which CNTs can be synthesized. We shall however restrict our scope to the use of pre-synthesized commercially available of SWNTs and using a solution suspension to make the device. Carboxylated-SWNTs (SWNT-COOH 80-90% purity) were dispersed (1 μg/mL) in dimethyl formamide. Though carbon nanotubes are known to have poor solubility in most solvents, however, the amide group of dimethyl formamide (DMF) can attach to the surface of the nanotubes making it a suitable solvent for obtaining uniformly suspended SWNTs in DMF [40]. The suspension of SWNTs in DMF solution was prepared in steps which involved first sonicating the dispersed carboxylated SWNTs followed by centrifugation at a speed of 10,000 rpm to remove soluble fraction and aggregates.

One of the issues in fabricating nanotube based sensor device is making attempts for their easy manipulation, which includes the controlled positioning and precise integration into electronic circuits. After the successful preparation of SWNT dispersion, they can be deposited on top of SiO2/Si or flexible substrate to fabricate FET. There are several methods by which this can be accomplished. Some of the widely used techniques are Langmuir-Blodgett deposition, Self-assembled monolayers (SAM)-assisted deposition, drop casting and dielectrophoresis alignment. Drop casting happens to be the simplest of these techniques. However, it curtails any control on the density of CNT network that is deposited across the electrodes.

Dielectrophoresis is a very versatile technique employed in aligning nanotubes across metal contacts [41]. The procedure involves addition of SWNTs suspension solution and subsequent application of an AC voltage at a predetermined frequency across the electrodes. When dielectric particles are exposed to a non-uniform electric field, charges move away from their initial balanced positions. The charge redistribution creates electric dipole moments that force the nanotube to rotate and position along the electric field lines. Therefore, the polarized nanotubes are subjected to a net force and can be aligned to follow the electric field direction (Figure 3) if they are free to move in a medium. Dielectrophoretic force field therefore self-limits the directed assembly to a single nanotube or nanotube bundle at predefined locations between the electrodes. The dielectrophoretic alignment of CNTs is affected by many factors including the dimension of the nanotubes, the properties of the medium, the strength of the electric field, bias voltage, frequency, deposition time, width of the electrodes and nanotube solution concentration. The aligned SWNTs are then annealed at an elevated temperature in an atmosphere of hydrogen and nitrogen gas. This is mostly done in order to improve the contact between the electrodes and the SWNTs and to remove DMF residues accumulated during the process of alignment.

2.1. Electropolymerization of CPs on SWNT Surface

CPs can be incorporated onto the surface of SWNTs either by possessing functional groups that are reactive toward SWNTs [42-44] or by in situ polymerization [45]. The former approach of functionalization limits its application pool by incorporating one kind of polymer or sometimes one sort of polymers on to the surface of SWNTs. In situ polymerization overcomes this issue as it enables the polymerization of a wide range of polymers on SWNTs based on doping type molecular interactions. Electropolymerization offers a simple, cost effective and controlled approach in depositing conducting polymer film on the surface of the CNTs. By altering parameters like charge, electrolyte composition and deposition potential, it imparts a great deal of flexibility in terms of controlling the thickness of the CP [22].

Electrochemical functionalization of SWNTs with the concerned CP can be conducted at ambient temperature with a three electrode configuration where the SWNTs network with the gold electrodes, a Pt wire, and a chlorinated silver wire (Ag/AgCl wire) are employed as working, counter, and reference electrodes, respectively. A suitable volume of the electrolyte solution i.e., the monomer and dopant is placed on top of the SWNT networks followed by potentiostatic electropolymerization. The electrodes are placed in contact with the electrolyte to constitute an electrochemical cell. Charge controlled electropolymerization was performed at different charges to find an optimum charge that illicits the maximum sensing response from the CNT-CP sensor.

3. Gas Sensing

For the purpose of gas sensing, the electrode pairs are wire bonded and packaged to an n-pin ceramic inline package, which is then mounted on to the breadboard of the sensing system set up. Real time measurements can be carried out by integrating the devices to a fully automated LabView set up. In a conventional gas sensing set up, a small volume of air tight sealed glass chamber with gas inlet and outlet ports for gas flow-through is positioned over the sensor chip. For sensing of volatile organic compounds, saturated vapors of the VOCs are produced by bubbling dry air through the bubbler column containing the VOC of interest. The carrier gas in this set up was dry air. Various concentrations of the vapors were introduced to the gas chamber by regulated flow of the vapors with the help of mass flow controllers. Prior to all the sensing experiments, the sensors are first exposed to dry air (purity: 99.998%) to attain a steady baseline resistance following which they were exposed to different concentrations of the saturated vapors of the VOC mixed in dry air for 15 minutes with an interval of 20 min recovery in air between two successive vapor exposures. This constitutes one complete cycle. This cycle is repeated for n number of times based on the range of concentrations being sensed. For real time gas sensing, the sensors are thus fabricated to be operated as chemiresistive sensors wherein the change in resistance upon exposure to an analyte forms the basis for measuring the sensitivity of the sensor. All sensors with an initial resistance in a limited range are used for sensing to ensure better reliability and consistency in gas sensing results.

Normalized resistance change is defined by

$$\Delta R/R_0\ (\%) = 100^*\ (R - R_0)/R_0,$$

where R is the resistance of the device upon analyte exposure and R_0 is the initial baseline resistance before analyte exposure.

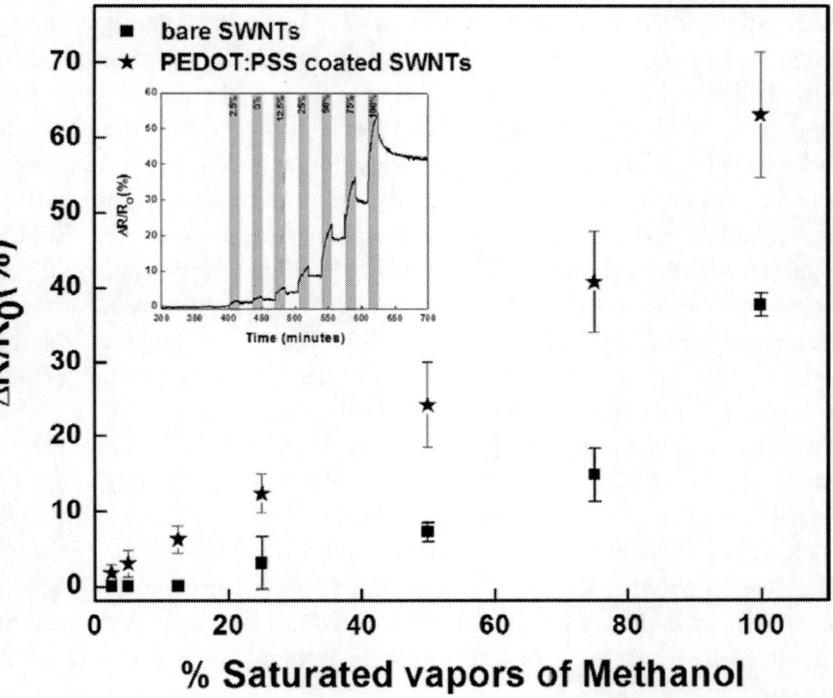

Figure 3a. Calibration curves for bare and PEDOT:PSS coated SWNTs device; dynamic response of the PEDOT:PSS coated SWNTs sensor towards different concentrations of methanol vapors (inset) [46].

3.1. Gas Sensing Using CNT-CP Hybrid Structures/Composites

In an attempt to demonstrate the gas sensing behavior of some of the CPs coated CNT devices, we recently reported the fabrication of PEDOT: PSS coated SWNT nano gas sensor [46].

Poly(3,4-ethylenedioxythiophene) doped with poly(styrene sulfonic acid) (PEDOT:PSS) is widely used in chemical sensors owing to its properties of being electrochemically active, environmentally stable and maintaining its activity over a wide pH range. The wide application pool of most conducting polymers and their derivatives is often limited due to their inability to be used in a broad pH range [47]. The use of PSS as the counter ion renders enhanced conductivity and facilitates easier processibility [48].

Following the experimental steps mentioned in the preceding section, surface functionalization of SWNTs with PEDOT: PSS using electropolymerization yielded devices which were then configured in the chemiresistive mode of sensor configuration for gas sensing. These sensors showed higher sensitivity towards vapours of three popular volatile organic compounds namely methanol, ethanol and methyl ethyl ketone whose concentration regulation is critical in industrial manufacturing processes. The dynamic response of the hybrid sensors towards methanol and the corresponding calibration curve are shown in Figure 3a. Structural characterization of the aligned SWNT devices is done using Scanning Electron Microscope (SEM) which is shown in Figure 3b. As can be seen, for the case of electropolymerized PEDOT: PSS, a thin film of the polymer coating is formed uniformly all over the SWNT surface. The combination of a low oxidation potential and relatively low band gap gives PEDOT its unique electrochemical and spectroscopic properties. Band gap, in case of PEDOT, is located at the transition between the visible and near-IR regions of the spectrum, thus making it strongly cathodically colouring and transmissive to visible light, sky blue transparent, in the doped and conducting state [49]. Upon electropolymerization in presence of polyelectrolyte PSS at room temperature, the change of the redox state leads to a change in the electronic structure, which can be also observed as an optical colour transition to dark blue in optical microscope image.

To gain an understanding underlying mechanism governing the interaction of the CP coated SWNT sensors with the analyte, the sensors are configured in the chemFET mode of operation with polysilicon as the back gate. For example, Figure 3c shows the transfer characteristics i.e., current vs. gate voltage (I_{DS}-V_{GS}) characteristics for the PEDOT: PSS coated SWNT FET device after exposure to air and then subsequent concentrations of vapors of VOC (in this case, methanol) at V_{DS} of -1 V. Upon exposure to increasing concentrations of the analyte, there occurs a change in the slope of the transfer characteristics curve along with a shift in the threshold voltage (V_{TH}) towards more negative voltage values of the gate voltage. V_{TH} is defined as the voltage required to deplete carriers in the channel.

It can be calculated by fitting the I_{DS}-V_{GS} curve with a tangent line at the point of maximum transconductance and determining the intercept of this line with the gate voltage (V_{GS}) axis. Methanol, in this case, being an electron-donating species, exposure to it causes a shift in the valence band away from the Fermi level, resulting in a decrease in the charge carrier (hole) concentration. This thereby causes a reduction in conductivity and a negative shift in the threshold voltage (V_{TH}) of the device. Thus, decrease in threshold voltage can be attributed to the electrostatic "chemical gating" of the device wherein adsorption of the analyte induces additional negative charge in the SWNT, thus n-doping the SWNT and shifting the I_{DS}-V_{GS} curve towards more negative gate voltages. In addition, there occurs a change in the mobilities (calculated from the device transconductance i.e., slope of the I_{DS} vs. V_{GS}) of the device when exposed to different saturated concentrations of methanol vapors. Thus, shift in threshold voltage along with change in mobility in the case of the device suggests that the sensing mechanism is governed both by the electrostatic gating and Schottky barrier effects in combination [50].

Conducting polymers such as PEDOT: PSS are soluble in water miscible solvents and high polar alcohols as well. Exposure to alcohols therefore results in the penetration of the vapor into the organic film and in the re-dissolution of the film at high vapor concentrations. This leads to loss of conduction paths in the polymer film thereby increasing the resistance [51]. Similar explanations to the increase in resistance upon vapor exposure have also been related to the swelling of the polymer which in turn increases the hopping distance for charge carriers [35], [52] and hence lowering of conductivity.

Choi et al. reported the fabrication of a gas sensor based on PANI) and SWNT composite materials and its sensing behavior towards NH_3 and CO [53]. In their work, PANI was synthesized through the chemical oxidative polymerization of aniline in an aqueous acidic media. PANI is one of the conducting polymers for gas sensing applications due to its ease of preparation, relatively high electrical/environmental stability and low resistivity [54, 55]. PANI/SWNTs films were prepared by drop casting the mixed solution of SWNTs and PANI on a substrate. The PANI/SWNTs composite material sensors showed a faster response to NH_3 gas than CO gas. It was observed that CO gas increased the conductance of the composite, while NH_3 gas showed an opposite response. The mechanism of sensing was explained in terms of CO gas molecules that were adsorbed chemically onto the surfaces of the PANI/SWNTs acted as hole donors, while the NH_3 molecules which were physically adsorbed acted as hole acceptors [56, 52]. Figure 3d shows the schematic of the developed sensor and the dynamic response of the sensors towards the gases under study.

Castro and co-workers have demonstrated an electronic nose (e-nose) consisting of hierarchically structured SWNT-conductive polymer nano composites (CPC) for detecting volatile organic biomarkers [57]. Each sensor comprises of CNTs and the spray layer by layer (sLbL) assembly of polymer matrices. The resultant e-nose could detect nine organic solvent vapours (isopropanol, tetrahydrofuran, dichloromethane, *n*-heptane, cyclohexane, methanol, ethanol, water and toluene) by pattern recognition tool PCA enabling separation of vapour clusters.

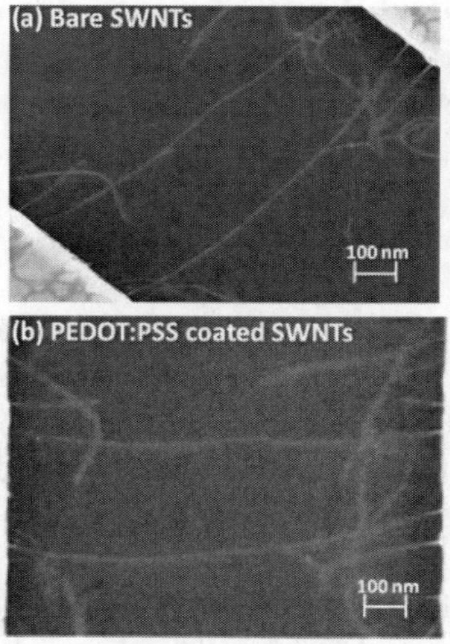

Figure 3b. SEM image of the same SWNT device before (a) and (b) after PEDOT:PSS electropolymerization [46].

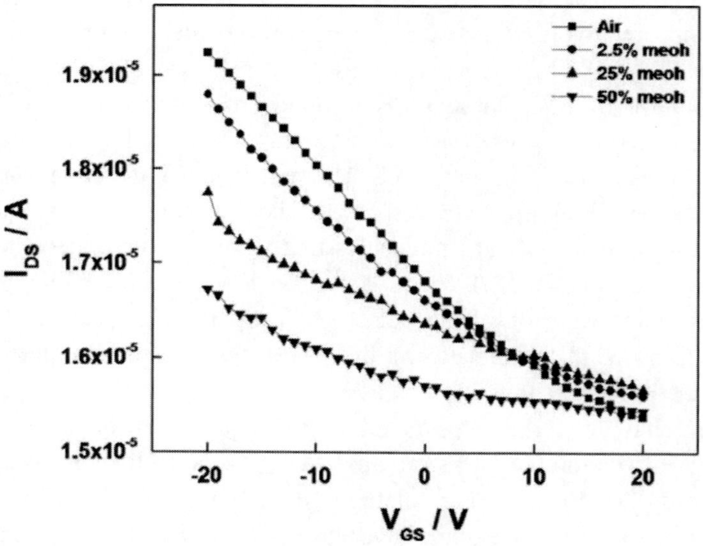

Figure 3c. Transfer characteristics curves of PEDOT:PSS coated SWNTs device when exposed to air and different saturated concentrations of methanol vapors performed at $V_{DS} = 1$ V [46].

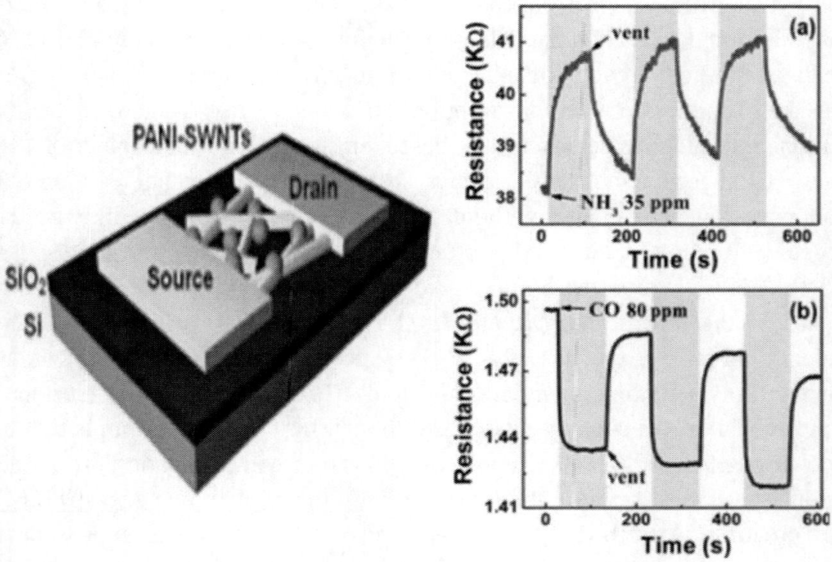

Figure 3d. Schematic of the PANI/SWNT composite (on left) and the response of the composite to NH_3 and CO (on right) [53] (with permission).

4. Biosensing Using CNT-CP Hybrid Structures/Composites

CPs have attracted significant interest in the development of biosensors which are being used widely in fields of biotechnology, food and agriculture product processing for detection of food borne pathogens [58, 59], health care and medicine [60, 61]. The electrically conducting polymers are known to possess numerous features in terms of stability and ease of processing. They can be synthesized by one-step electrosynthesis in the form of an adherent film deposited at the surface of the electrode with anionic species being included by doping. This doping reaction makes it possible to modulate the conductivity reversibly over several orders of magnitude via redox interactions. All these features allow them to act as excellent materials for immobilization of biomolecules and rapid electron transfer for the fabrication of efficient biosensors. Of the several generations of biosensors categorized depending upon their mode of operation, CP based biosensors come under the category of direct binding of the biocatalyst to an electronic device that transduces and amplifies the signal. Reports demonstrating use of electropolymerized polypyrrole films as electrochemical sensing devices for glucose sensors, both enzymatic and non-enzymatic have been found vastly in literature [62, 63]. Realizing the advantages of nano arrays in terms of miniaturization, greater sensitivity and less power consumption, efforts were also made to successfully demonstrate glucose sensor arrays based on polyaniline nano junctions [64]. These CP based nano sensor arrays were very specific and exhibited fast responses toward glucose without being affected by factors like need for oxygen and redox mediators that often plague conventional enzymatic sensors.

Recently we reported boronic acid-substituted polyaniline coated SWNTs sensor for the detection of saccharides [65]. The sensors were fabricated using dielectrophoretic alignment of SWNTs followed by electropolymerization of the CP Poly (3-aminophenylboronic acid) (PABA) on SWNTs surface by experimental procedure discussed in the earlier section. Complexation of saccharides (as well as alkyl and aromatic diols) with aromatic boronic acids produces a stable boronate anion and a proton in the pH range of 6–10 [66, 67] thus creating possibilities for development of electrochemical approaches for sugar detection. Functionalization of SWNTs by aromatic boronic acid can be performed either by covalent attachment of 3-APBA or non-covalent attachment routes. In this work, the non-covalent approach of attachment by the electropolymerization of 3-APBA in the presence of fluoride is performed. This high concentration of fluoride resulted in sustained polymerization and significant and continuous polymer growth.

Upon exposure to saccharide, the presence of fluoride enhances the complexation of the saccharides with the BA receptors by disrupting any B–N interactions and facilitates the formation of the anionic ester [68]. This ester formation leads to an increase in the relative negative charge on the surface of the p-type SWNT channel, thereby resulting in an increase in the resistance of the sensor. Thus, the extent of the binding event is detected and quantified by the PABA-coated SWNT chemiresistive sensor. Unlike the enzymatic assays which consume the analyte in the course of the detection, this approach utilizes the pH dependent reversible nature of boronic acid–saccharide complexation wherein the sugar is released back into the solution at lower pH values. The schematics of the mechanism of detection and the response of the sensor towards D-fructose at neutral pH are shown in Figure 4a.

Dai and co-workers have synthesized conducting polymer-SWNT coaxial nanowires by electrochemically depositing a concentric layer of polypyrrole uniformly onto each constituent aligned CNT (inset of Figure 4b) [69]. Further, they demonstrated these CP-CNT coaxial nanowires for the amperometric detection of glucose as shown in Figure 4b [70]. Glucose oxidase was immobilized aligned carbon nanotube substrate by electropolymerization of pyrrole in the presence of glucose oxidase. The glucose oxidase-containing polypyrrole carbon nanotube coaxial nanowires were used to monitor the concentration change of hydrogen peroxide (H_2O_2) generated from the glucose oxidation reaction by measuring the increase in the electro-oxidation current at the oxidative potential of H_2O_2 (i.e. the amperometric method) as shown in Figure 4b. The amperometric response was found to be much higher than that of more conventional flat electrodes coated with glucose oxidase containing polypyrrole films under the same conditions.

As shown in Figure 4b, a linear response of the electrooxidation current to the glucose concentration was obtained for the CP-NT nanowire sensor. The coaxial structure allows the nanotube framework to provide mechanical stability and efficient thermal/electrical conduction to and from the conducting polymer layer thereby making these electrodes very attractive for sensing applications.

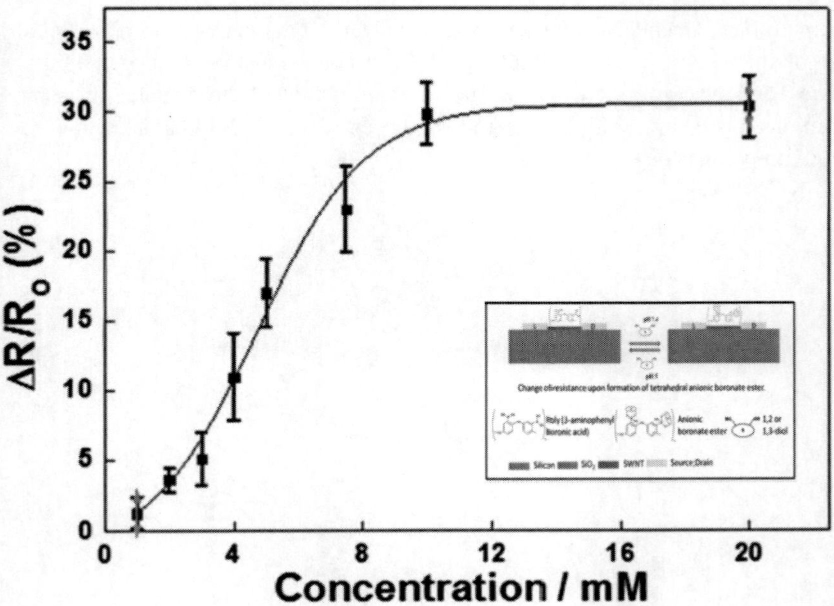

Figure 4a. Response of the PABA-coated SWNT sensors towards D-fructose at pH of 7.4 with the corresponding sigmoidal fit. (inset) Schematic of the chemiresistive PABA-coated SWNT sensor [65].

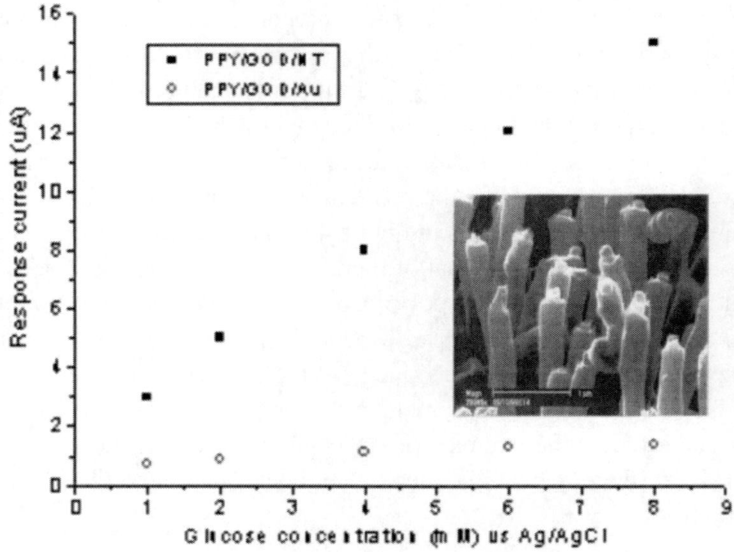

Figure 4b. The dependence of electro-oxidation current at the oxidative potential of H_2O_2 on the glucose concentration for the CP-NT coaxial nanowire sensor (solid squares) and the conventional polypyrrole sensor on a flat electrode under the same conditions (open circles) [70]. (inset) SEM images of the CP-NT coaxial nanowires [69] (with permission).

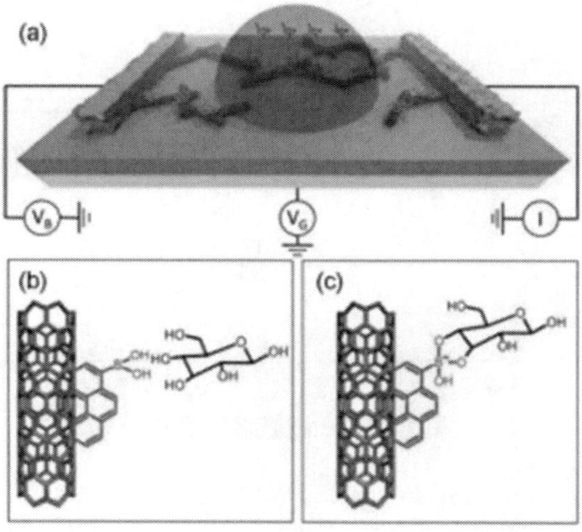

Figure 4c. a) Schematic of the experimental geometry. b) Illustration of glucose binding to a nanotube functionalized with pyrene-1-boronic acid. c) Bound glucose forms a boronate anion complex that has electrostatic effects on the nanotube FET [71] (with permission).

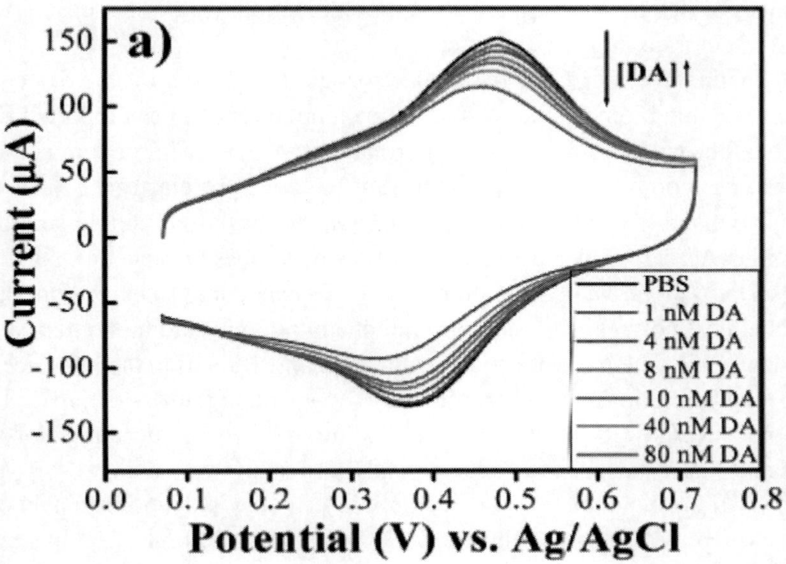

Figure 4d. Cyclic voltammograms of the PABA/ssDNA/single-walled nanotube (SWNT) nanocomposite in PBS and in the presence of different concentrations of dopamine [73] (with permission).

In an attempt to develop a non-invasive solution for detection of D-glucose in human body. Lerner and coworkers recently reported a scalable, label-free all-electronic D-glucose sensor based on carbon nanotube transistor functionalized with pyrene-1-boronic acid [71]. This sensor responds to glucose in the range 1 μM – 100 mM, which includes typical glucose concentrations in human blood and saliva. The FET-based detection technique involved drop-casting a high density of CNTs (150–300 nanotubes in the channel) for sensor fabrication and covalent functionalization of pyrene-1-boronic acid on CNTs. The principle of operation is illustrated in Figure 4c. The sensor exhibited a high sensitivity and selectivity towards D-glucose and a lower limit of detection (LOD) of 300 nM.

Other related work in the field of using boronic acid receptor modified CNT based electrochemical sensors involve CNT sensor integrated in a microfluidic channel for the detection of sugars using a sensitive impedance-based detection technique [72]. The BA receptors were coupled to the surface of the aligned carbon nanotubes through electrochemical functionalization using both types of chemical coupling i.e., covalent and non-covalent. The mechanism of detection is explained on a combination of affinity sensing and the field effect transistor (FET) effect.

Conducting polymer-CNT based electrochemical sensors for the detection of diol containing biomolecules such as dopamine have also been reported [73, 74]. Phenylboronic acid forms stable boronate ester bond with dopamine, and the resulting complex can be employed to develop electrochemical or electronic sensors for dopamine detection. On the basis that single-stranded DNA (ssDNA) can disperse bundled carbon nanotubes in aqueous solution PABA/ssDNA/single-walled nanotube (SWNT) nanocomposites by "in situ" electrochemical polymerization of 3-aminophenylboronic acid monomers were synthesized [75]. The response of this composite in PBS and in the presence of varying concentrations of dopamine is shown in Figure 4d [73]. The nanocomposite showed excellent properties through synergistic effects of the component materials. The electrocatalytic reductive ability of the ssDNA/SWNTs and their strong interaction with the polyaniline backbone significantly improved the stability of PABA. Further, combining the high performance of the developed poly (anilineboronic acid)/carbon nanotube composite with the excellent perm selectivity of an ion-exchange polymer Nafion, the detection of dopamine could be achieved without any interference from ascorbic acid.

CONCLUSION

In conclusion, conducting polymer-carbon nanotubes hybrid structures provide an ideal platform for building reliable and highly sensitive sensors for a wide array of sensing applications [76, 77]. The combination of CNT (with their electronic and mechanical properties) and CP (with its sensing behavior) offers an attractive route not only to reinforce the polymer film but also to introduce a variety of new electronic properties based on morphological modifications or electronic interactions between the two components giving a robust blend of optimum properties that can be exploited to develop superior electrochemical sensors.

In this chapter, we have focused our attention mainly on research findings related to CNT-CP hybrids from our group along with some other significant work from literature. Even this short chapter reveals the unlimited scope of CNT-CP hybrids/composites in sensing applications employing different detection techniques. Research on the interface between nanomaterials and conducting polymers is emerging as one of broadest and most dynamic areas of science and technology as it addresses sensing needs with high sensitivity and selectivity. With so many surface functionalization techniques already developed and advanced device fabrication methods being developed, it creates endless opportunities to address challenges in the field of medical diagnostics using CNT-CP hybrids based lab-on-a-chip systems and e-nose systems for gas/vapour sensing for environmental monitoring and air pollution detection and control.

ACKNOWLEDGMENTS

We acknowledge the assistance of Prof. Nosang V. Myung, Dr. Chaker Tlili and Dr. Rajat K. Paul.

REFERENCES

[1] Goldstein, N. (1997). *Handbook of Nanophase Materials*, New York: Dekker.

[2] Iijima, S. (1991). *Nature*, *354*, 56.

[3] Tans, S. J., Verschueren, A. R. M. & Dekker. C. (1998). *Nature*, *393*, 49–52.

[4] Martel, R., Schmidt, T., Shea, H. R., Hertel, T. & Avouris, P. (1998). *Appl. Phys. Lett.*, *73*, 2447–2449.

[5] Yao, Z., Postma, H. W. C., Balents, L. & Dekker, C. (1999). *Nature*, *402*, 273–276.

[6] Baughman, R. H., Cui, C. X., Zakhidov, A. A., Iqbal, Z., Barisci, J. N., Spinks, G. M., Wallace, G. G., Mazzoldi, A., DeRossi, D., Rinzler, A. G., Jaschinski, O., Roth, S. & Kertesz, M. (1999). *Science*, *284*, 1340–1344.

[7] Bondavalli, P., Legagneux, P. & Pribat, D. (2009). *Sensors and Actuators,* B, *140*, 304–318.

[8] Justin Gooding, J., Wibowo, R., Jingquan, W., Yang, D., Losic, S., Orbons, F. J. Mearns, J. G. & Shapter, D. B. (2003). Hibbert. *Journal of the American Chemical Society*, *125*(30), 9006-9007

[9] Yu, X., Chattopadhyay, D., Galeska, I., Papadimitrakopoulos, F. & Rusling, J. F. (2003). *Electrochemistry Communications*, *5*(5), 408-411.

[10] Patolsky, F., Weizmann, Y. & Prof. I. W. (2005). *Angewandte Chemie-International Edition*, *43*(16), 2113-2117.

[11] Guiseppi-Elie, A., Lei, C. & Baughman, R. H. (2002). *Nanotechnology*, *13*(5), 559-564.

[12] Chen, R. J., Bangsaruntip, S., Drouvalakis, K. A., Kam, N. W. S., Shim, M., Li, Y., Kim, W., Utz, P. J. & Dai. H. (2003). *Proceedings of the National Academy of Sciences of the United States of America*, *100*(9), 4984-4989.

[13] Park, D. W., Kim, Y. H., Kim, B. S., So, H. M., Won, K., Lee, J. O., Kong, K. J. & Chang. H. (2006). *Journal of Nanoscience and Nanotechnology*, *6*(11), 3499-3502.

[14] Cella, L. N., Sanchez, P., Zhong, W., Myung, N. V., Chen, W. & Mulchandani, A. (2010). *Analytical Chemistry*, *82*(5), 2042-2047.

[15] Cella, L. N., Chen, W., Myung, N. V. & Mulchandani, A. (2010). *Journal of the American Chemical Society*, *132*(14), 5024-+.

[16] Kong, J., Franklin, N. R., Zhou, C., Chapline, M. G., Peng, S., Cho, K. & Dai, H. (2000). *Science*, *287*(5453), 622-625.

[17] Someya, T., Small, J., Kim, P., Nuckolls, C. &Yardley, J. T. (2003). *Nano Letters*, *3*(7), 877-881.

[18] Chung, J., Lee, K., Lee, J., Troya, D. & Schatz, G. C. (2004). *Nanotechnology*, *15*(11), 1596-1602.

[19] Srivastava, S., Sharma, S. S., Kumar, S., Agrawal, S., Singh, M. & Vijay, Y. K. (2009). Int. *J. Hydrog. Energy*, *34*, 8444–8450.

[20] Huyen, D. N., Chien. N. D. & Korean Phys, J., (2008). *Soc*. *52*, 1364–1367.

[21] Star, A., Joshi, V., Skarupo, S., Thomas, D. & Gabriel, J. C. P. (2006). *J. Phys. Chem. B*, *110*, (2006) 21014–21020.

[22] Mubeen, S., Zhang, T., Yoo, B., Deshusses, M. A. & Myung, N. V. (2007). *J. Phys. Chem.* C, *111*, 6321–6327.

[23] Kauffman, D. R., Sorescu, D. C., Schofield, D. P., Allen, B. L. & Jordan, K. D. (2010). *A. Star. Nano Lett*. *10*, 958–963.

[24] Mubeen, S., Zhang, T., Chartuprayoon, N., Rheem, Y., Mulchandani, A., Myung, N. V. & Deshusses, M. A. (2010). *Anal. Chem*., *82*, 250–257.

[25] Wongwiriyapan, W., Inoue, S., Okabayashi, Y., Ito, T., Shimazaki, R., Maekawa, T., Suzuki, K., Ishikawa, H., Honda, S., Mori, H., Oura, K. & Katayama, M. (2009). *Appl. Phys. Express*, *2*, 095008–095011.

[26] Kumar, M. & Castro, J. F. (2012). Feller, *Carbon*, (2012) *50*, 3627–3634.

[27] Zilberman, Y., Tisch, U., Shuster, G., Pisula, W., Feng, X., M¨ullen, K. & Haick, H. (2010). *Adv. Mater.*, *22*, 4317–4320.
[28] Wang, Y., Xiong, H., Gao, Y. & Li, H. (2008). *J. Mater. Sci.*, *43*, 5609–5617.
[29] Heeger, A. J. (2001). *Angewandte Chemie International Edition*, *40*(14), 2591-2611.
[30] MacDiarmid. A. G. (2001). *Angewandte Chemie International Edition*, 40(14), 2581–2590.
[31] McQuade, D. T., Pullen, A. E. & Swager, T. M. (2000). *Chem Rev*, *100*(7), 2537-2574.
[32] Wallace, G. G. & Kane-Maguire, L. A. P. (2002). *Adv. Mater.*, *14*(13-14), 953-960.
[33] Garnier, F. (1989). *Angewandte Chemie-International Edition in English, 28*(4), 513-517.
[34] Hangarter, C. M., Bangar, M., Mulchandani, A. & Myung, N. V. (2010). *Journal of Materials Chemistry*, *20*(16), 3131-3140.
[35] Cao, Y., Kovalev, A. E., Xiao, R., Kim, J., Mayer, T. S. & Mallouk, T. E. (2008). *Nano Letters*, *8*(12), 4653-4658.
[36] Svetlicic, V., Schmidt, A. J. & Miller, L. L. (1998). *Chemistry of Materials*, *10*(11), 3305-+.
[37] Tan, C. K. & Blackwood, D. J. (2000). *Sensors and Actuators B-Chemical*, *71*(3), 184-191.
[38] Liu, H., Kameoka, J., Czaplewski, D. A. & Craighead. H. G. (2004). *Nanoletters*, *4*, *671-675.*
[39] Dan, Y. P., Cao, Y. Y., Mallouk, T. E., Evoy, S. & Johnson, A. T. C. (2009). *Nanotechnology*, *20*, 434014–434018.
[40] Kong, J., Franklin, N. R., Zhou, C. W., Chapline, M. G., Peng, S., Cho, K. J. & Dai, H. J. (2000). *Science*, *287*, 622–625.
[41] Dan, Y. P., Cao, Y. Y., Mallouk, T. E., Johnson, A. T. & Evoy, S. (2007). *Sens. Actuators B – Chem.*, *125*, 55–59.
[42] Liu, Y. Q., Yao, Z. L. & Adronov, A. (2005). *Macromolecules*, *38*, 1172-1179.
[43] Liu, Y. L. & Chen, W. H. (2007). *Macromolecules*, 40, 8881-8886.
[44] Liu, Y. L., Chang, Y. H. & Liang. Polymer, M. (2008). *49*, 5405-5409.
[45] Lee, K. P., Showkat, A. M., Gopalan, A. I., Kim, S. H., Choi, S. H. & Sohn, S. H. (2006). *Journal of Applied Polymer Science*, *101*, 3721-3729.
[46] Badhulika, S., Myung, N. V. & Mulchandani, A. (2014). *Talanta*, *123*, 109–114.

[47] Nikolou, M. & Malliaras, G. G. (2008). *Chem. Rec.*, *8*, 13–22.
[48] Jonas, F., Krafft, W. & Muys, B. (1995). *Macromol. Symp.*, *100*, 169–173.
[49] Groenendaal, L., Jonas, F., Freitag, D., Pielartzik, H. & Reynolds, J. R. (2000). *Adv. Mater.*, *12*, 481–494.
[50] Heller, A. M., Janssens, J., Mannik, E. D., Minot, S. G. & Lemay, C. (2008). *Dekker. Nano Lett.*, *8*, 591-595.
[51] Mabrook, M. F., Pearson, C. & Petty, M. C. (2005). *Applied Physics Letters*, *86*, pp.
[52] Bai, H. & Shi, G. Q. (2007). *Sensors*, *7*, 267-307.
[53] Choi, H. H., Lee, J., Dong, K., Ju, B. & Lee, W. (2012). *Macromolecular Research*, *20*, 2, 143-146.
[54] Ramanathan, K., Bangar, M. A., Yun, M. H., Chen, W. F., Mulchandani, A. & Myung, N. V. (2004). *Nano Lett.*, *4*, 1237.
[55] Wang, J., Chan, S., Carlson, R. R., Luo, Y., Ge, G. L., Ries, R. S., Heath, J. R. & Tseng, H. R. (2004). *Nano Lett.*, *4*, 1693.
[56] Zhao, J., Buldum, A., Han, J. & Lu, P. (2002). *Nanotechnology*, *13*, 195.
[57] Castro, M., Kumar, B., Feller, J. F., Haddi, Z., Amari, A. & Bouchikhi, B. (2011). *Sensors and Actuators B: Chemical.*, *159*, 1, 213-219.
[58] Minett, A. I., Barisci, J. N. & Wallace, G. G. (2003). *Analytica Chimica Acta*, *475*(1-2), 37-45.
[59] Arshak, K., Velusamy, V., Korostynska, O., Oliwa-Stasiak, K. & Adley, C. (2009). *IEEE Sensors Journal*, *9*(12), 1942-1951.
[60] Kajiya, Y., Sugai, H., Iwakura, C. & Yoneyama, H. (1991). *Analytical Chemistry*, *63*(1), 49-54.
[61] Kumar, A., Rajesh, A., Chaubey, S. K. & Grover, B. D. (2001). *Malhotra. Journal of Applied Polymer Science*, 82(14), 3486-3491.
[62] Bidan, G. (1992). *Sensors and Actuators B-Chemical*, *6*(1-3), 45-56.
[63] Zotti, G. (1992). *Synthetic Metals*, *51*(1-3), 373-382.
[64] Forzani, E. S., Zhang, H., Nagahara, L. A., Amlani, I., Tsui, R. & Tao, N. (2004). *Nano Letters*, *4*(12), 2519-2519.
[65] Badhulika, S., Tlili, C. & Mulchandani, A. (2014). *Analyst*, *139*, 3077.
[66] Springsteen, G. & Wang, B. H. (2002). *Tetrahedron*, *58*(26), 5291–5300.
[67] Liu, L., Xia, N., Xing, Y. & Deng, D. (2013). *Int. J. Electrochem. Sci.*, *8*(9), 11161–11174.
[68] Westmark, P. R., Valencia, L. S. & Smith, B. D. (1994). *J. Chromatogr. A*, *664*(1), 123–128.

[69] Gao, M., Huang, S., Dai, L., Wallace, G., Gao, R. & Wang, Z. (2000). *Angew. Chem. Int. Edn Engl.*, *39*, 3664.

[70] Gao, M., Dai, L. & Wallace, G. (2003). *Electranayl.*, *11*, 1089.

[71] Lerner, M. B., Kybert, N., Mendoza, R., Villechenon, R., Lopez, M. A. B., Johnson, A. T. C. (2013). *Appl. Phys. Lett.*, *102*, 183113.

[72] Vlandas, A., Kurkina, T., Ahmad, A., Kern, K. & Balasubramanian, K. (2010). *Anal. Chem.*, *82*(14), 6090–6097.

[73] Ali, S. R., Ma, Y., Parajuli, R. R., Balogun, Y., Lai, W. Y. C. & He, H. (2007). *Anal. Chem.*, *79*, 2583–2587.

[74] Wu, W., Zhu, H., Fan, L., Liu, D., Renneberg, R. & Yang, S. (2007). *Chem. Commun.*, 2345–2347.

[75] Ma, Y. F., Ali, S. R., Dodoo, A. S. & He, H. X. (2006). *J. Phys. Chem. B*, *110*, 16359-16365.

[76] Dai, L., He, P. & Li, S. (2003). *Nanotechnology*, *14*, 1081–1097.

[77] Feller, J. F., Castro, M. & Kumar, B. (2011). Edited by T McNally, Queen's University Belfast, UK and P Pötschke, Leibniz-Institut für Polymerforschung Dresden e.V. (Leibniz Institute of Polymer Research Dresden), Germany, ISBN 1 84569 761 8, ISBN-13: 978 1 84569 761 7, Q1, 750 pages.

INDEX

C

D

E

F

G

H

I

J

K

L

M

N

O

P

Q

R

S

T

U

V

W

X

Y